The Infinite Pickle

Puzzles for Kids, Educators, Parents and Puzzlers

This book is for sharing!

Download it for free on MathPickle.com.

Go to www.Patreon.com/MathPickle to support me in creating the next book.

First published by Our Street Books, 2024
Our Street Books is an imprint of Collective Ink Ltd.,
Unit 11, Shepperton House, 89 Shepperton Road, London, N1 3DF office@collectiveinkbooks.com
www.collectiveinkbooks.com
www.our-street-books.com

For distributor details and how to order please visit the 'Ordering' section on our website.

ISBN: 978 1 80341 685 4

A CIP catalogue record for this book is available from the British Library.

Illustrations by Okan Bulbul

Book design by Camila Macca

Printed in China

The Infinite Pickle

Most puzzles require general rules and unique hints.

Examples of general rules:

- In sudoku, only one number goes in each square.
- In crossword puzzles, words are horizontal or vertical, not diagonal.

Examples of unique hints:

- Today's sudoku puzzle has the number 6 in the top left square.
- Today's crossword puzzle has the hint "26 Across: Greek god of hunting."

Infinite Pickles are different. They only require general rules. These rules are sufficient to create an infinite number of puzzles from simple to complex.

Because their difficulty can be dialed up or down, **Infinite Pickles** can be used to engage struggling students and gifted students simultaneously. The educator does not need to prepare multiple activities to fit different students. These are one-pickle-fits-all activities. 🙂

The adult puzzler will also enjoy the increasingly difficult challenges that **Infinite Pickles** offer. Challenge yourself, compete against me, or get together and compete against your friends & enemies.

For each **Infinite Pickle**, I've included some of my best answers. I often do not spend the time to prove that these are the best answers—and I also don't use computers to help me. That means you may beat me. That's fun! It already happened between the first and second printing. Amelie Cao, a sixth-grade girl from New York, beat my high score on one puzzle, with a score of 2133. I'll leave you to find which puzzle!

Whenever some of my best answers are coming up, I'll yell **"SPOILER ALERT!"**

Table of puzzles

I like each and every chapter I read (and I did read ALL). I truly LOVE the *Coloring Ukraine* puzzle (the map of Ukraine may be changing, unfortunately...) and also the *Bad Rulers* one [from the sequel: *The Infinite +1 Pickle]*. Wow, this is some combinatorics made accessible to elementary school kids.

Nitsa Movshovitz-Hadar
Haifa, Israel

Nitsa is a Professor Emerita of mathematics education at Technion in Haifa.

***The Infinite Pickle* ... Brilliant!!! I'm dumbstruck by how creative and innovative these puzzles are. Wowza!**

James Tanton
Arizona, USA

James is the "Mathematician-at-Large" for the Mathematical Association of America. He is the inventor of exploding dots.

Google "Exploding dots"

Thank you so much for these fantastic puzzles and games! They give so much joy!

Gabriella Pinter
Milwaukee, USA

Gabriella is a professor in the Department of Mathematical Sciences, University of Wisconsin-Milwaukee.

Just like an Indian meal is never complete without a pickle, a mathematical foundation is never complete without the playfulness exemplified in *The Infinite Pickle.* Every chapter comes with fun, simple, colorful, and interesting games and activities which tickle the gray cells. This is the best way to learn mathematics—through pure fun and play.

Kiran Bacche
Bangalore, India

Kiran runs the Bangalore Math Circle. He is co-founder of dhimath.org.

WHERE DID PAGE 5 GO?

Most books use a predictably boring sequence to number pages. **What a waste!**

Before you flip each page, try to figure out the number on the following page. It is the next integer that is the sum of two previous page numbers in exactly one way.

You can see a list of all the Ulam numbers under 1000 on page 847.

Thanks to Stanisław Ulam of Lviv, Ukraine for this sequence!

Apples for when you work with kids!

Too often, those of you working with kids have started teaching math armed only with algorithms-to-be-mastered and pedagogic techniques to coerce compliance. Let's start afresh. Knowing some great puzzles can make your time spent doing math with little people joyous and productive.

Just as you can't be a great English teacher using Harlequin Romances, you cannot be a great math educator using mediocre puzzles. The best puzzles help with the most important things that we want for our kids. In order of importance we want them to:

#1 problem-solve,
#2 persevere after a failure,
#3 gain mathematical skills.

That's a sufficient list for parents and tutors, but teachers also need to:

#4 engage struggling and gifted students simultaneously.

The **Infinite Pickles** in this book do all of these things.

With an **Infinite Pickle,** your objective is not to teach things in as EASY a way as possible, but to teach things in as ENGAGING a way as possible. It is EASY to teach the algorithm of addition. The trouble is before you open your mouth the top 20% of your class already knows how to add; the bottom 20% of your class is struggling to remember last year's curriculum.

An **Infinite Pickle** allows you to ENGAGINGLY keep the whole class together until 90% understand how it works. When the class then breaks off into pairs, you are free to circulate—helping a pair group where they most need it. Top kids are deflected into problem solving (never a waste of time) and struggling kids work on skill acquisition.

Working with kids is an experimental science. Some of my ideas will work for you—some will not. Discard and adopt ideas as you see fit. I wish you all the best.

Gord!

PS. Throughout the book I'll be using the apple when I have something to say to those of you working with kic🙂

When teaching that 29 is less than 30, replace a normal lesson with Uncut Spaghetti. When teaching counting, replace a normal lesson with Jumping Frogs. When teaching place value, use Coloring Ukraine. **All of these Infinite Pickles can be used instead of standard classroom experiences. They fit into any curriculum.** They can also be used with much older children because all of them celebrate problem solving, which is the core of a quality mathematics education.

Pickles for when you geek out!

You feel like a geek today. It's time to jump into a beautiful puzzle. This book will give you tons to choose from. When you're ready to check your answers, you can compare them to my best attempts.

When you're finished self-indulging your puzzle-brain it might be time to help out others. Most of these puzzles would traditionally be categorized as recreational mathematics. That trivializes them. Their more esteemed claim to fame is that they are pedagogic gems. They engage the full spectrum of student ability.

Some of you were gifted children. Your teacher had to find a way to inspire you while not losing the struggling child seated next to you. This is impossible using mediocre problems. The real testament of these puzzles is not that you would have found them engaging but that all of your peers would have found them engaging too.

If you start interacting with elementary school classrooms, you will learn as much as you will give. The subtleties of inspiring young people may not be your strength, but there is much value in this collaboration.

You might think your best chance for meaningful outreach with young people is in high school, but in my opinion the most effective use of your outreach time is working side-by-side elementary school teachers. Here you will find students who are inquisitive, teachers who are open to collaboration and a curriculum that is more flexible than in high school.

Gord!

PS. Throughout the book I'll be using the pickle when I have something to say to those geeking out. 🙂

Uncut Spaghetti

I once visited a miserly puzzler who had numbered dinner plates. When I became hungry, a single spaghettum was boiled up.

The puzzler asked me to choose a number. **I chose 14.** My spaghhetum was then laid out on my plate—starting at 14 and always going to the smallest neighbouring number that had not yet been visited.

× Diagonal moves are not allowed.

The miserly puzzler said that I could eat my spaghettum if I filled the plate—otherwise the spaghettum would need to be cut and that was no good. **I would be left without supper.**

- Was 14 a good choice?
- What numbers work?
- What numbers fail?
- Are there any patterns that will persist for plates of different sizes?

SPOILER ALERT!

0	1	2	3	4	5
6	7	8	9	10	11
12	13	14	15	16	17
18	19	20	21	22	23
24	25	26	27	28	29
30	31	32	33	34	35

0	1	2	3	4	5
6	7	8	9	10	11
12	13	14	15	16	17
18	19	20	21	22	23
24	25	26	27	28	29
30	31	32	33	34	35

Yes—14 was a good choice. All of the spaghettum got onto the plate without needing to be cut. We are going to color 14 green to celebrate.

Not all numbers work. If 31 had been chosen, the spaghettum would not fill the plate. I'd need to cut it. As all Italians know, a cut spaghettum does not taste as good as an uncut spaghettum.

I'll color 31 red as a warning!

On the following page we will show the red-green coloring of this 6x6 plate and other plates ranging from 1x1 to 8x8.

SPOILER ALERT!

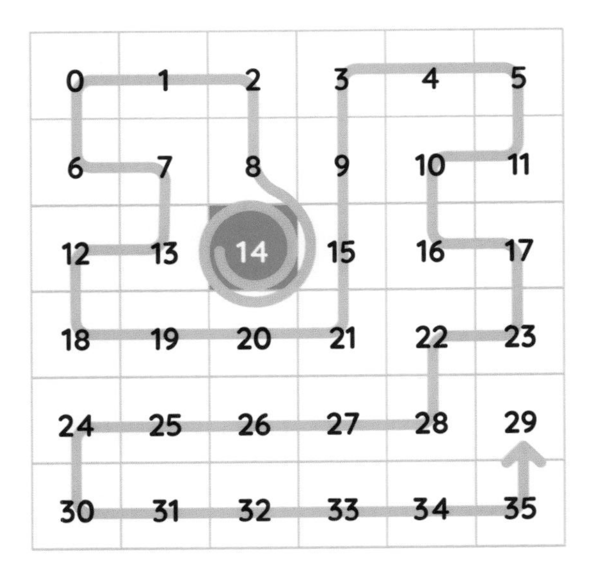

0	1	2	3	4	5
6	7	8	9	10	11
12	13	14	15	16	17
18	19	20	21	22	23
24	25	26	27	28	29
30	31	32	33	34	35

All classrooms have children who struggle with self esteem. What question can you ask if you want to guarantee a child with low self esteem a positive experience when you call on them in front of the whole class?

"Connor—Choose a square that you think will fail."

This is good pedagogy for young children because if Connor finds a square that fails, you can celebrate. If Connor finds a square that succeeds—you can celebrate in a different way. In both cases you aim to bypass the self-esteem issues so the child can relax and enjoy getting into the puzzle.

0	1
2	3

0	1	2	3
4	5	6	7
8	9	10	11
12	13	14	15

0	1	2	3	4	5
6	7	8	9	10	11
12	13	14	15	16	17
18	19	20	21	22	23
24	25	26	27	28	29
30	31	32	33	34	35

My 6x6 plate colored.

0	1	2	3	4	5	6	7
8	9	10	11	12	13	14	15
16	17	18	19	20	21	22	23
24	25	26	27	28	29	30	31
32	33	34	35	36	37	38	39
40	41	42	43	44	45	46	47
48	49	50	51	52	53	54	55
56	57	58	59	60	61	62	63

0

0	1	2
3	4	5
6	7	8

0	1	2	3	4
5	6	7	8	9
10	11	12	13	14
15	16	17	18	19
20	21	22	23	24

0	1	2	3	4	5	6
7	8	9	10	11	12	13
14	15	16	17	18	19	20
21	22	23	24	25	26	27
28	29	30	31	32	33	34
35	36	37	38	39	40	41
42	43	44	45	46	47	48

I chose this puzzle because I like eating spaghetti :) I like drawing the long spaghetti and then painting with green. I don't like when the spaghetti is cut short and I have to paint with red.

Arven just finished 1st grade, she loves math, but she loves doing art more. She lives in Ankara, Turkey.

Her mother, **Özgül,** has a PhD in Special Education.

Based on the patterns of plates 1-8, conjecture about the coloring of this 10x10 plate before seeing the solution on the next page.

Use a 1-100 wall chart if you can't find a 0-99 wall chart.

I use a 0-99 wall chart. These are difficult to find, but are superior to the more common 1-100 wall charts. Why?

- Because we should be identifying simple patterns and one of the most important for students is base ten.
- The numbers in each row have the same number of "tens."
- From left to right the "units" increase from 0-9.

I sometimes use a flip chart with numbers 100-199 on the second page and 200-299 on the third page, etc. Each page consists of the numbers that have the same number of "hundreds." If I stab at "72" I know that "172" is right underneath.

0	1	2	3	4	5	6	7	8	9
10	11	12	13	14	15	16	17	18	19
20	21	22	23	24	25	26	27	28	29
30	31	32	33	34	35	36	37	38	39
40	41	42	43	44	45	46	47	48	49
50	51	52	53	54	55	56	57	58	59
60	61	62	63	64	65	66	67	68	69
70	71	72	73	74	75	76	77	78	79
80	81	82	83	84	85	86	87	88	89
90	91	92	93	94	95	96	97	98	99

There are lots of patterns that you might have observed on the smaller plates:

- The column on the right seems to alternate green and red.
- The second column on the right seems to be mostly red.
- There is a checkerboard pattern covering part of the plate.
- The bottom row seems to be mostly red.

There is a difference in plates that have an odd and even edge length. This 10x10 plate has more in common with 6x6 and 8x8 plates than 5x5 and 7x7 plates.

No class should be expected to get all of the points above. Let the students conjecture and then check to see if they are right. Reward wrong conjectures even more than correct ones. That's how you inspire open curiosity and remove the stigma of failure.

0	1	2	3	4	5	6	7	8	9
10	11	12	13	14	15	16	17	18	19
20	21	22	23	24	25	26	27	28	29
30	31	32	33	34	35	36	37	38	39
40	41	42	43	44	45	46	47	48	49
50	51	52	53	54	55	56	57	58	59
60	61	62	63	64	65	66	67	68	69
70	71	72	73	74	75	76	77	78	79
80	81	82	83	84	85	86	87	88	89
90	91	92	93	94	95	96	97	98	99

On a 5x5 plate, add five meatballs so that all the other numbers can be colored green. (The spaghettum ignores all numbers covered with a meatball.)

SPOILER ALERT!

On a 4x4 plate, rearrange the numbers 0-15 so that you can color all numbers green.

SPOILER ALERT!

When a wet spaghettum hardens it might be strong enough to build things. Add numbers 1-11 to the circles on the tower and wheel. Make sure that no matter which circle you start at, moving repeatedly to the smallest connected number will let you visit all the circles.

SPOILER ALERT!

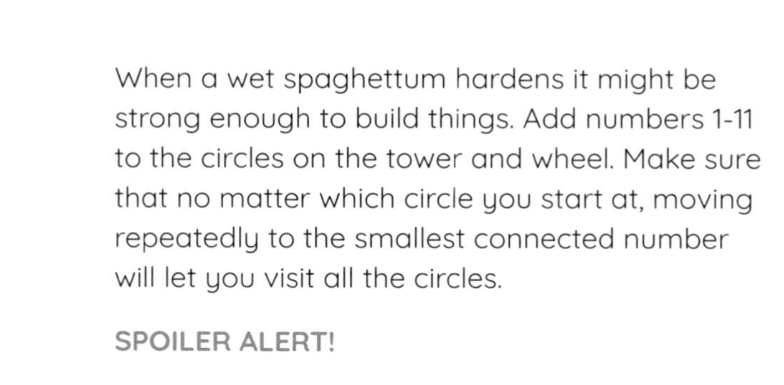

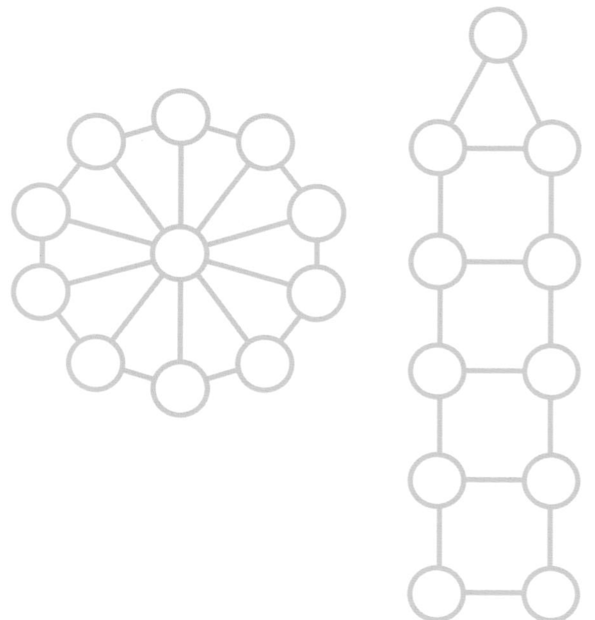

0	1	2	3	4
5	6	7	8	9
10	11	12	13	14
15	16	17	18	19
20	21	22	23	24

Five meatballs can be placed on a 5x5 plate so that all the other numbers are green.

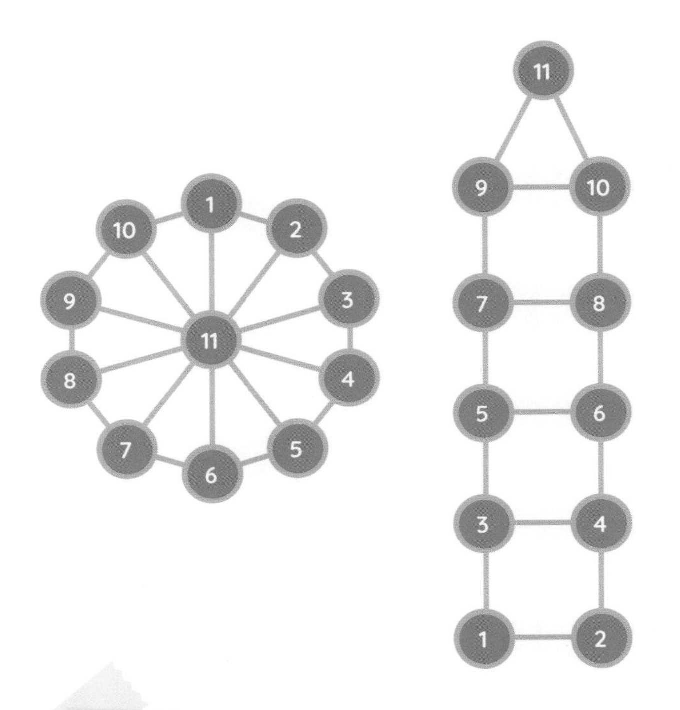

0	6	7	1
5	13	14	8
4	12	15	9
3	11	10	2

There are many ways to make the 0-15 plate work. Put the numbers 0-3 in the corner squares and put the numbers 12-15 in the center squares. Fill in the rest of the plate in any way.

If you want a beautiful online experience of plates up to size 20x20, google "data genetics uncut spaghetti."

Google "data genetics uncut spaghetti"

At all stages of solving symmetric puzzles students should be encouraged to think about the symmetry.

- If the puzzles above were too scary looking, a student should first try first solving a shorter tower or a wheel with fewer spokes.
- After a solution is found, a student should wonder if they have found a general solution that solves an infinite number puzzles—towers at any height and wheels with any number of spokes.

Complete numbering this house and mansion so they can remain all-green.

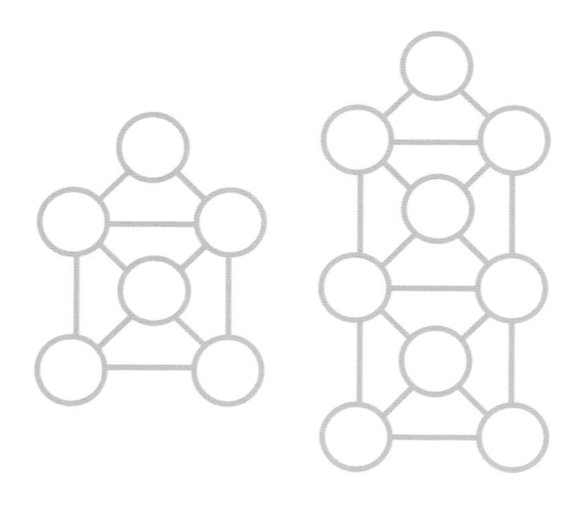

Harder: Add numbers 1-10 to the circles below. Make sure that no matter which circle you start at, moving repeatedly to the smallest connected number will let you visit all the circles.

SPOILER ALERT!

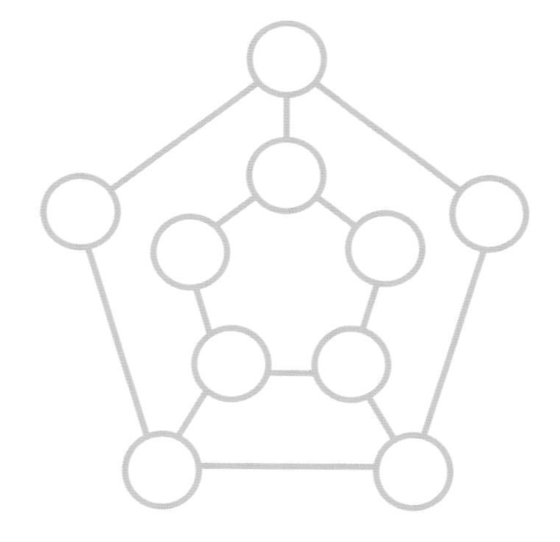

Put the positive integers 1-n on the n vertices of a graph. Color a vertex green if a path following the Uncut Spaghetti algorithm visits all vertices. Otherwise color the vertex red.

Which graphs can be painted all green by careful placement of the numbers? Can graphs with a loop through all vertices always be colored green? If not, find such a graph with fewest nodes or the fewest edges for which an all-green coloring does not exist. This 14-node graph looks like a good candidate, but I'm not confident:

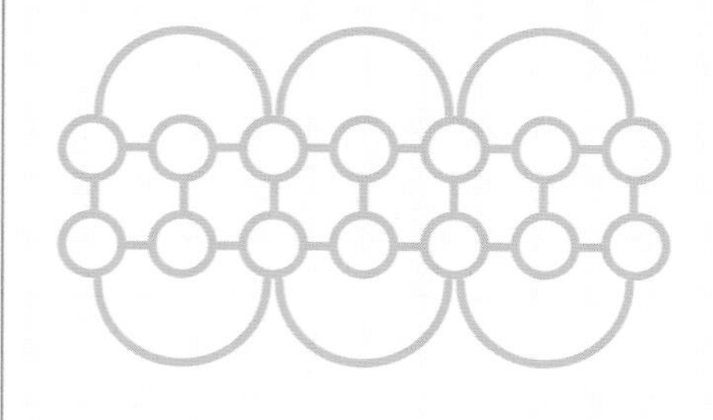

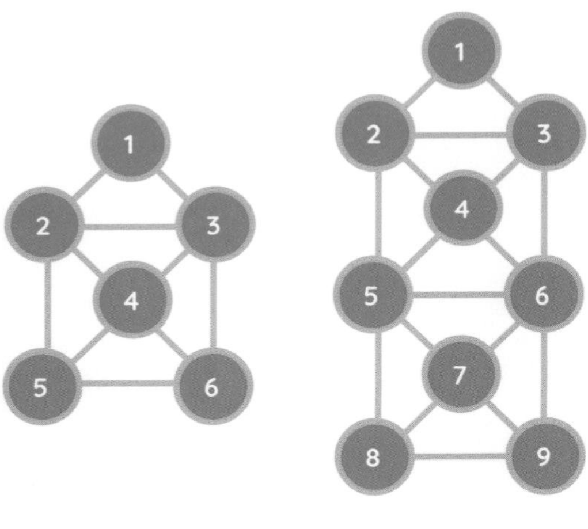

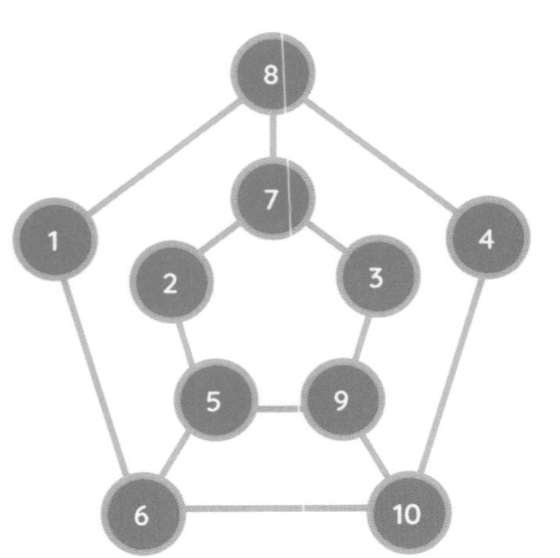

Let your students make their own puzzles. To begin, limit the maximum number of circles to 8. If you don't set some limit, the puzzles might be too difficult to figure out if they are possible or impossible.

Encourage students to explore classes of symmetric puzzles. These big puzzles at the right look beautiful and have solutions that might be generalized.

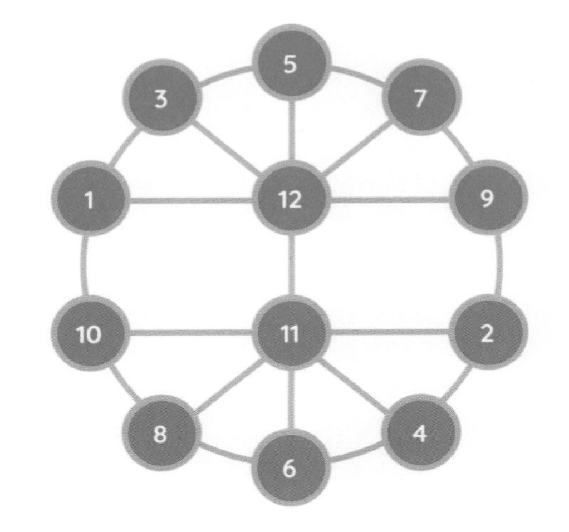

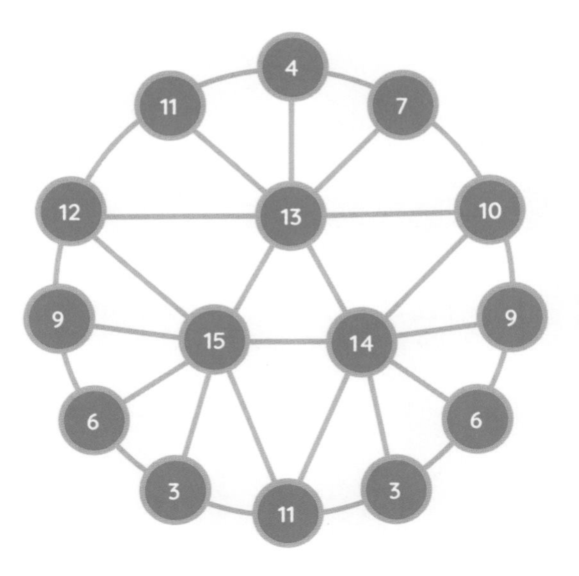

These string-art graphs can be numbered going clockwise. Some of them turn out to be all green. Experiment with other string-art graphs to see when they are all green.

Did you notice that all of these graphs have colorings that have mirror symmetry? Will this be true for larger loops?

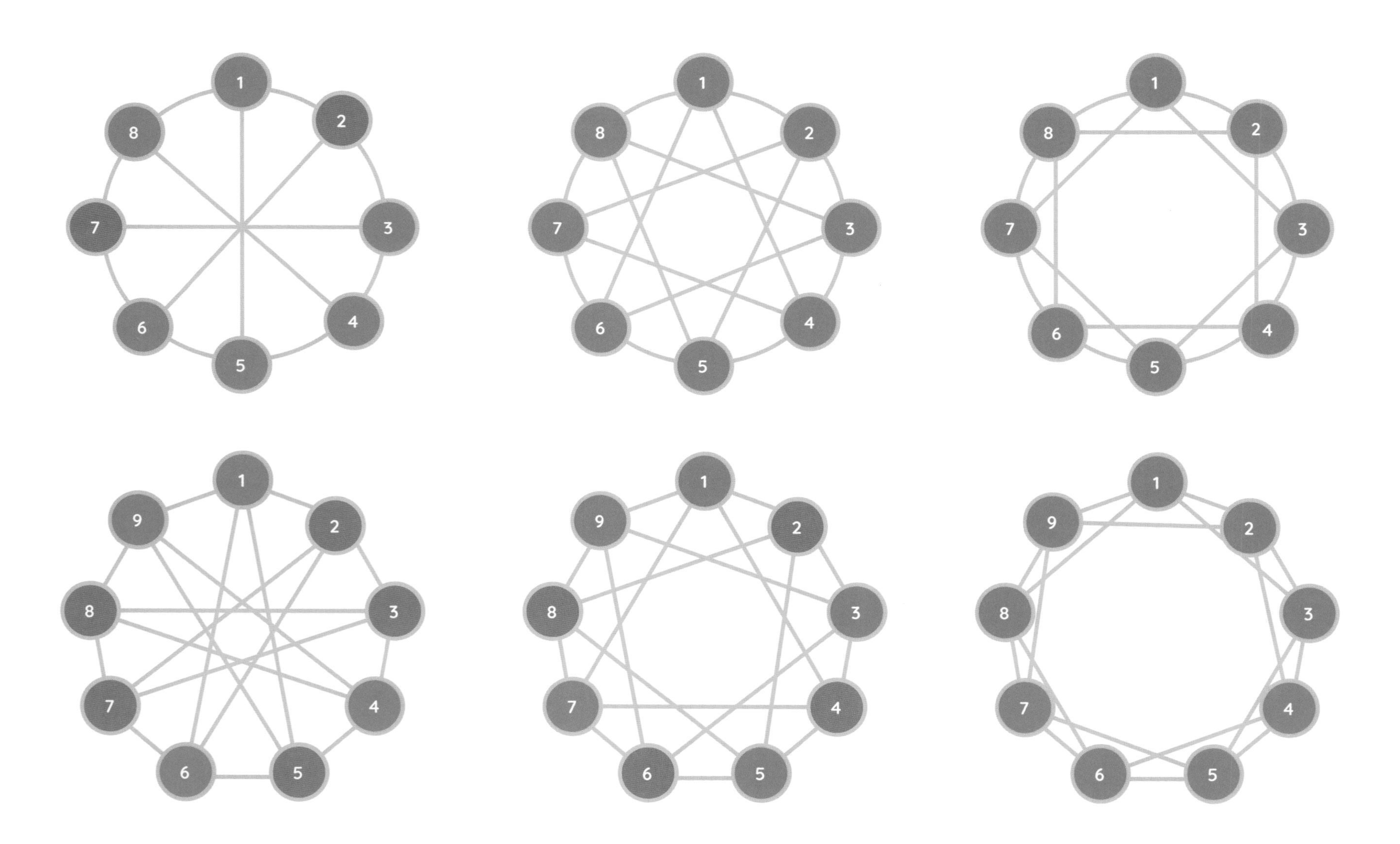

This confused me when I first explored it. The symmetry seemed to hold, but I couldn't think of why. Sure enough, it does fail. **Experiment with larger loops.**

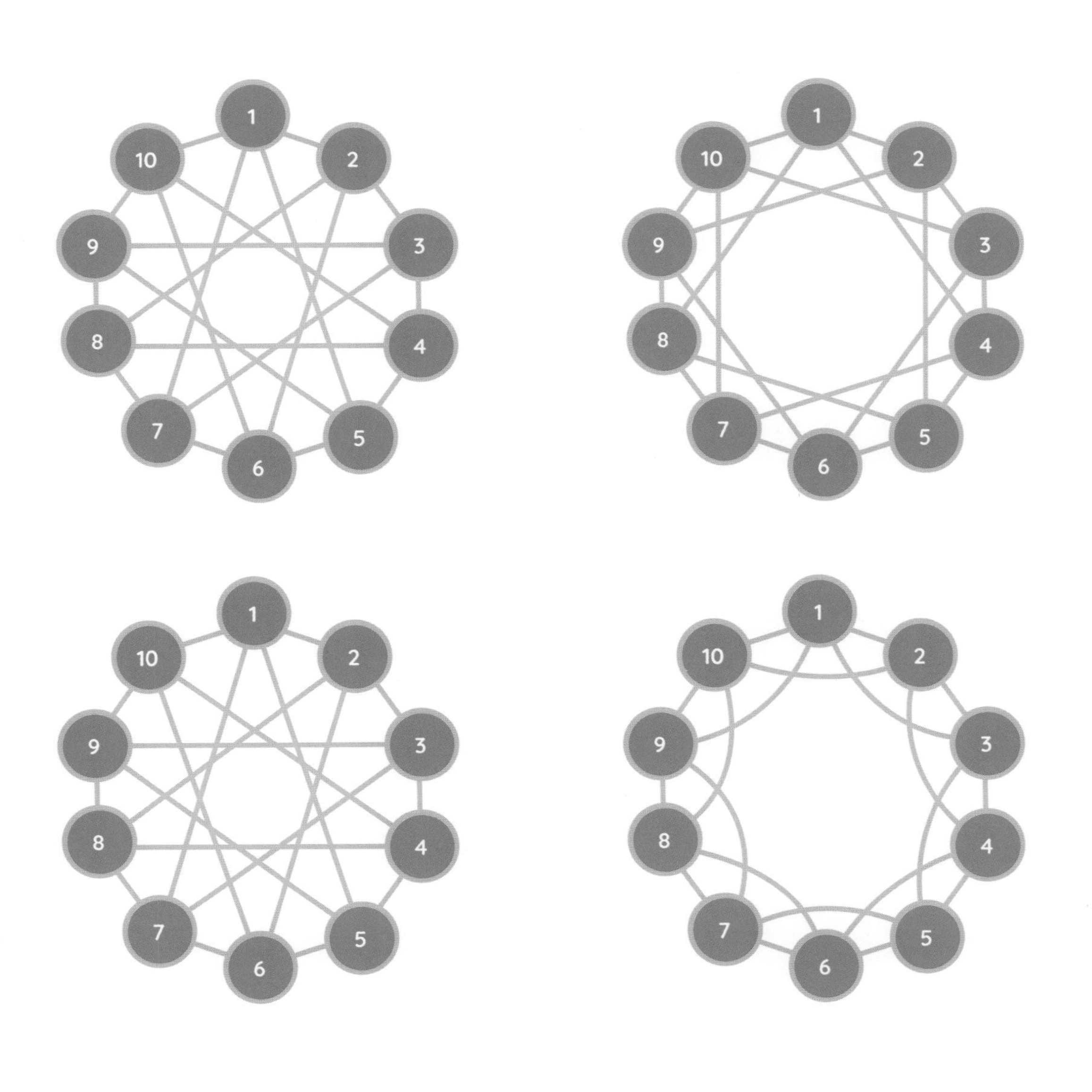

JUMPING FROGS

Whenever frogs are by themselves,
they play a game.

The frogs line up their lily pads. Each turn, all the frogs on one lily pad jump left or right. How far they jump depends on how many frogs are on the lily pad.

- one frog jumps one.
- two frogs jump two.
- three frogs jump three.
- etc.

The jumping frogs can never land on an empty lily pad.

Can all the frogs end up partying on the same lily pad?

This can be done for five frogs. **Try it!**

Don't follow the example of these 5 delinquent frogs. All their jumps are correct, but the results are sad.

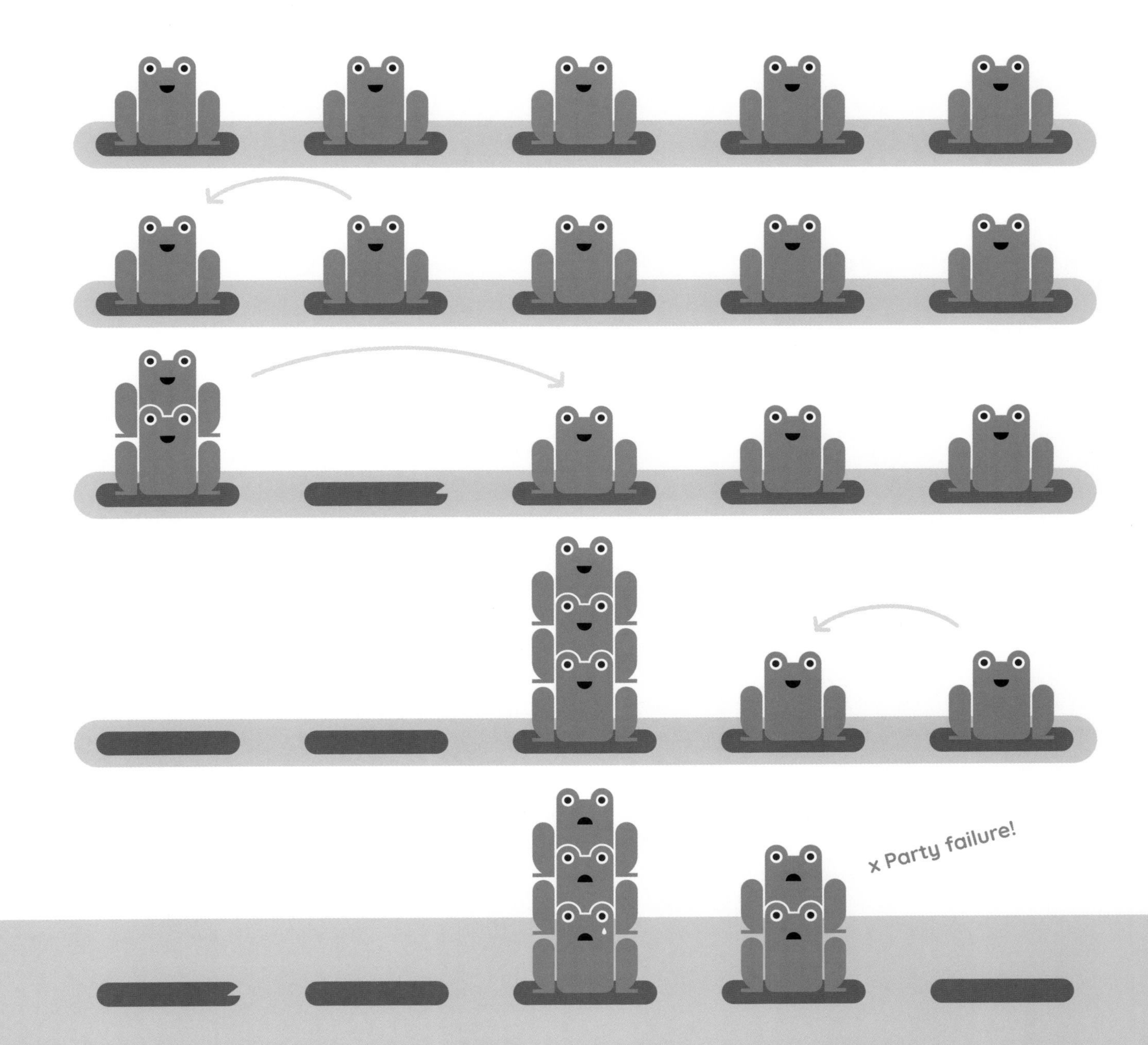

After solving it for five frogs, try 6, 7, 8 or 9 frogs in a line? **SPOILER ALERT!**

7 and 9 frogs can party together. In fact any odd numbers of frogs can party together. Here's how:

- Start in the middle.
- Take turns jumping left and right.

Melissa Raskauskas labels the lily pads with letters so her students can say things like: "Jump Frog C to the left." Labeling with numbers 1-10 can be confusing. Melissa's letters are better.

A similar back-and-forth jumping pattern will allow you to solve for all even numbers as well.

The queen frog shows up one night. She must be on the top of any pile she's in. **Can she party with all these frogs if she starts on lily pad 8?**

SPOILER ALERT!

Students love suggesting their own rules for puzzles. Try some with the whole class? Are they good? Are they terrible because they make the puzzle too simple or too hard to solve? Even if the immediate classroom results are mediocre—getting students to realize that rules are created by people like them—that's good pedagogy.

Here is one way to help the queen. The frog on lily pad 3 starts jumping back and forth and then our queen on lily pad 8 starts jumping back and forth. The big party happens on lily pad 5.

No matter where the queen begins, a jumping pattern similar to this ensures the frogs can always have their party!

French frogs abolished monarchy only to find themselves in just as awkward a pickle with new populist rules. Here is one: **"It is hereby declared that no pile of frogs shall jump on a smaller pile of frogs."**

Show how it is still possible for 13 French frogs to party.

SPOILER ALERT!

Although this new populist rule makes organizing parties more difficult, all numbers of frogs are still possible. This is possible to prove by junior high students.

SPOILER ALERT!

Two insights that helped me:

1. Any 2^n line of frogs can have a party in any location.
2. Any line of frogs can be broken up into powers-of-two groups.

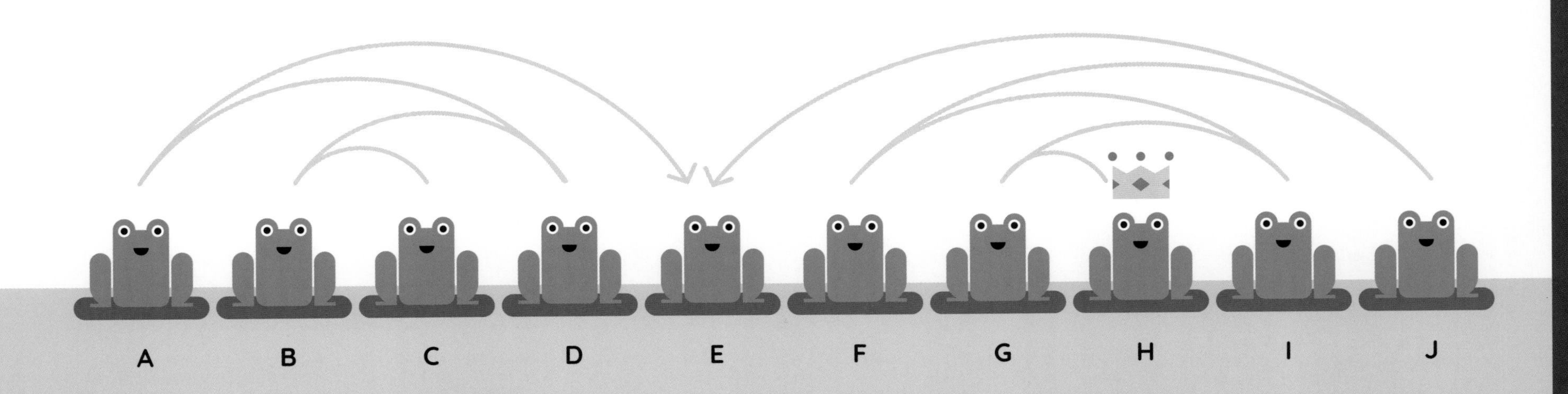

There are lots of solutions for the 13 lucky frogs. Here is one. Instead of showing everything, I'll just show the first nine jumps.

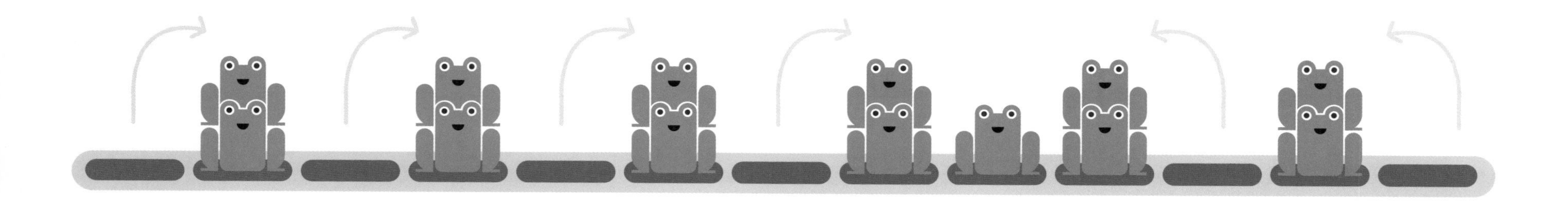

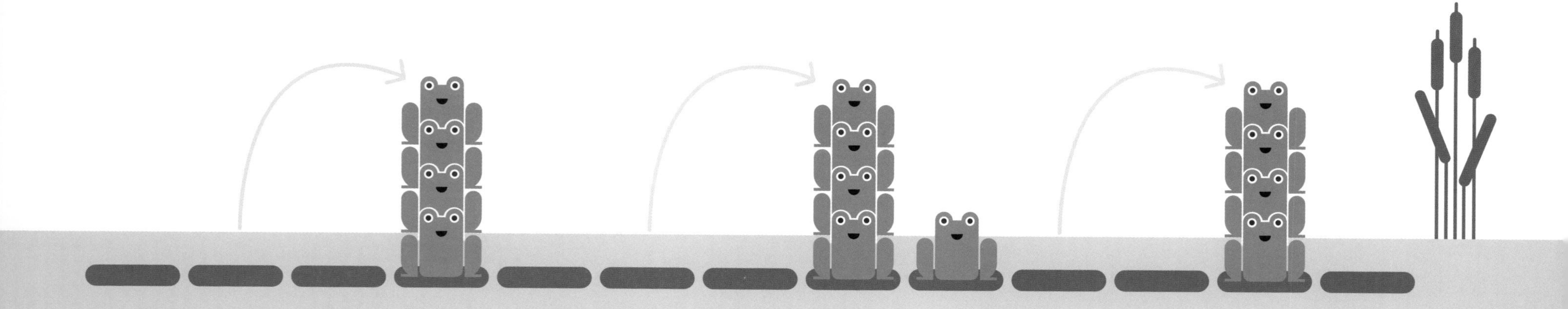

Let's go back to the original simple rules where it is ok for a large pile to jump on a small pile.

If the frogs make poor choices on their first jumps—possibly because they drank too much swamp water—it can mean that they are guaranteed to fail.

For 3-12 frogs what are the fewest number of bad jumps that will result in guaranteed failure?

SPOILER ALERT!

For three frogs there is no way to mess up. That was a trick question. 🙂

Give trick questions to students as often as possible. Try to keep a straight face. This keeps the classroom laughably fresh.

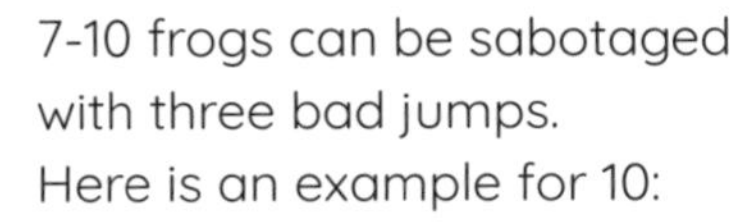

For 4-6 frogs the possibility of a party on a lily pad can be ruined with just two bad jumps. Here is an example with six:

7-10 frogs can be sabotaged with three bad jumps. Here is an example for 10:

I don't think there is a 3-move that sabotages eleven or more frogs. A 4-move sabotage seems possible up to about 25 frogs.

A little courtship game: before their life-changing kiss, the human princess and her frog-prince play a game.

- The princess wants the fewest possible number of parties.
- The frog-prince wants the most.

Starting with the princess, they alternate. On their turn they command the frogs on one lily pad to jump. Continue until there are no longer any jumps possible.

Start with five or six frogs. If both princess and frog-prince play well, how many parties (occupied lily pads) can we expect?

SPOILER ALERT!

With 5 frogs the princess will get her way with one giant party. With 6 frogs the princess and frog-prince will end up with two parties. Below is midway through a 10-frog game.

After the jumps below, it is the princess' turn. What jumps should she consider?

SPOILER ALERT!

On the next page, the pink jumps are the ones the princess should consider. Any of them will result in two parties instead of three.

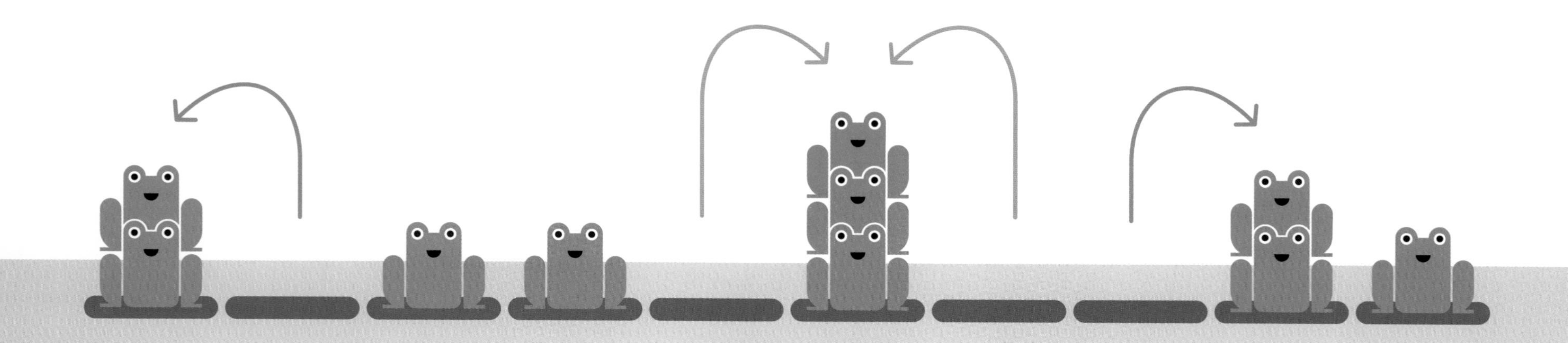

Sahiba is a high school student in Michigan, USA. She has hosted national mathematics festivals with play-based puzzles. Her love for mathematics grew from math competitions, puzzles and overall having fun with problem solving.

One night some of the frogs are no-shows. These party poopers can make it a lot more difficult or impossible for one big party to happen. The frogs below can do it, but it requires much froggy thinking.

SPOILER ALERT!

Google "Numberphile Frog Jumping"

This puzzle could be encoded in binary as 101011111111. That's not beautiful. Instead, frogs say that the fewest number of trailing ones that are required so that 1010 is solvable is eight.

The frogs like to figure out the minimum number of trailing ones for any binary number. For many the answer is zero. **These numbers have a special place in every froggy heart because it means a party can still go ahead if there are no-shows!**

The sequence of the smallest binary numbers that have an increasing number of trailing ones start like this: 0, 101, 1001, 1010. These can be solved by adding 0, 1, 2, and 8 trailing ones respectively. These are evil sequences to the frogs so out of respect I won't talk about them anymore. That's convenient since I don't know what the next terms are.

These jumps give a successful party for ten frogs. Jump 3 is not shown but you can probably figure out where it happened.

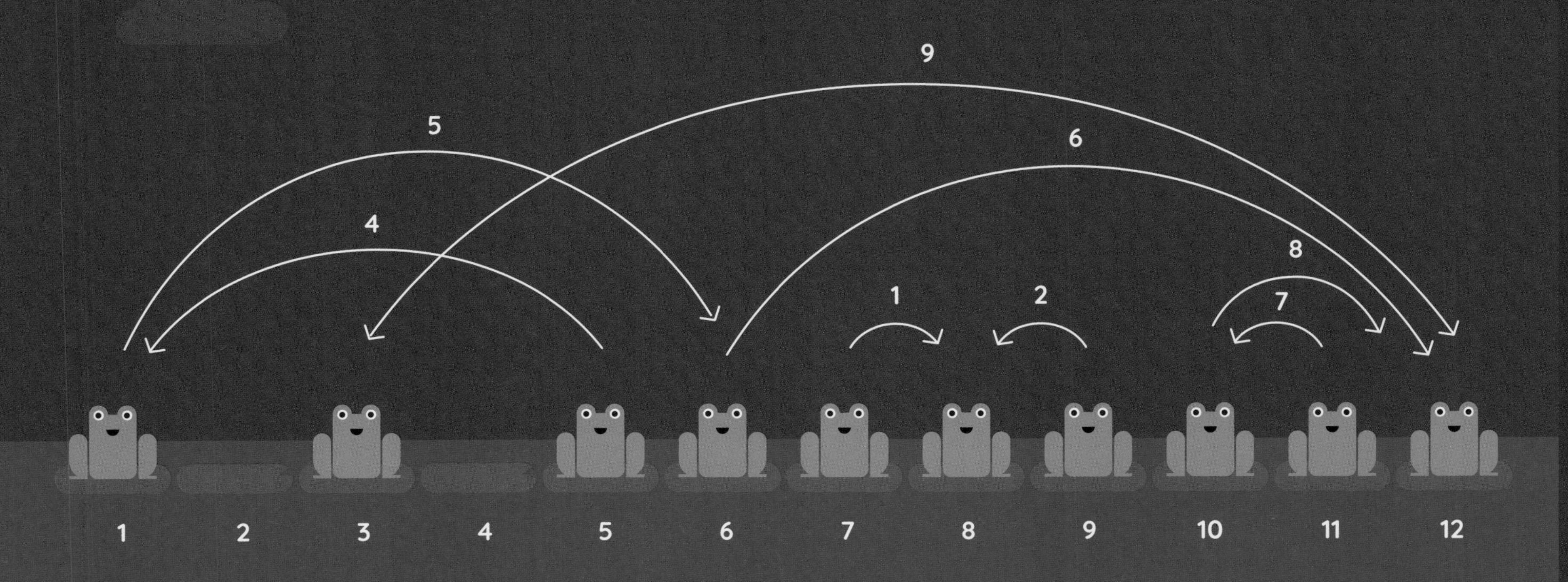

Some frogs have tired of the game of arranging their lily pads in a line. Instead, they are trying something new with new rules:

- One frog jumps one, two frogs jump two, etc.
- Frogs cannot jump onto an empty lilypad.
- All the lily pads must be linked together.
- When counting jumping distance, use connected lily pads.

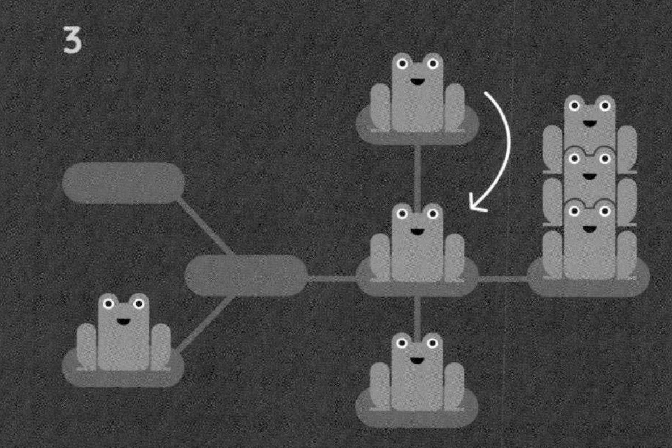

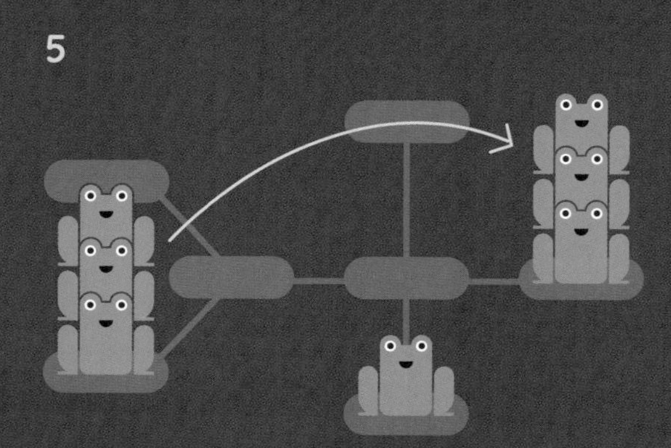

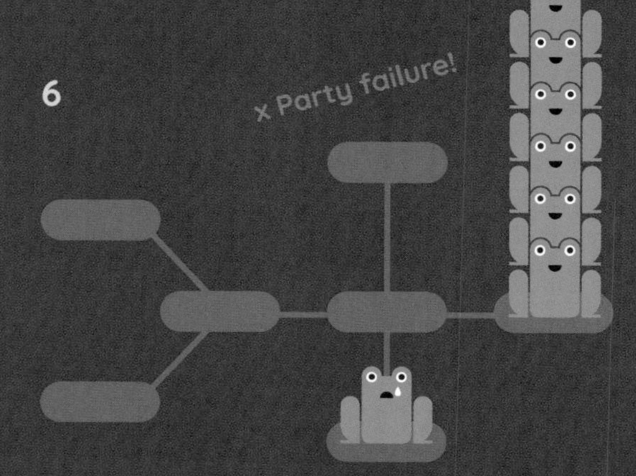

For this arrangement of seven lily pads, **which ones can end up hosting a successful party?**

Some arrangements of lily pads cannot have a successful party anywhere. Find an arrangement with the fewest lily pads for which a successful party is impossible.

SPOILER ALERT!

In this arrangement, there are only two lily pads where parties can occur. **Let's color them green.**

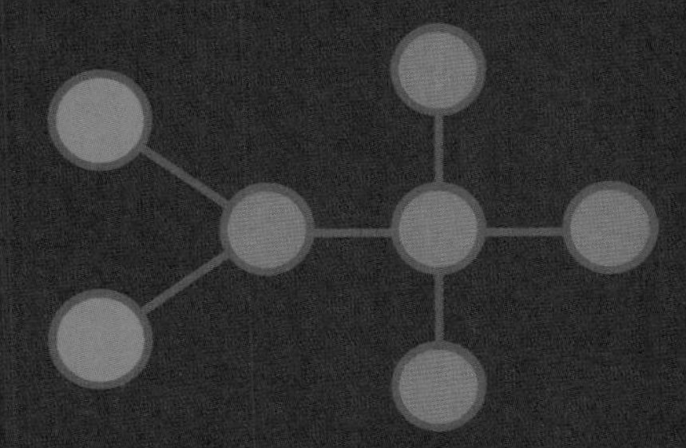

All arrangements of seven lily pads have at least one lily pad that can host a successful party. That's true of all arrangements of fewer lily pads as well.

Most arrangements of eight lily pads also work. Here are four:

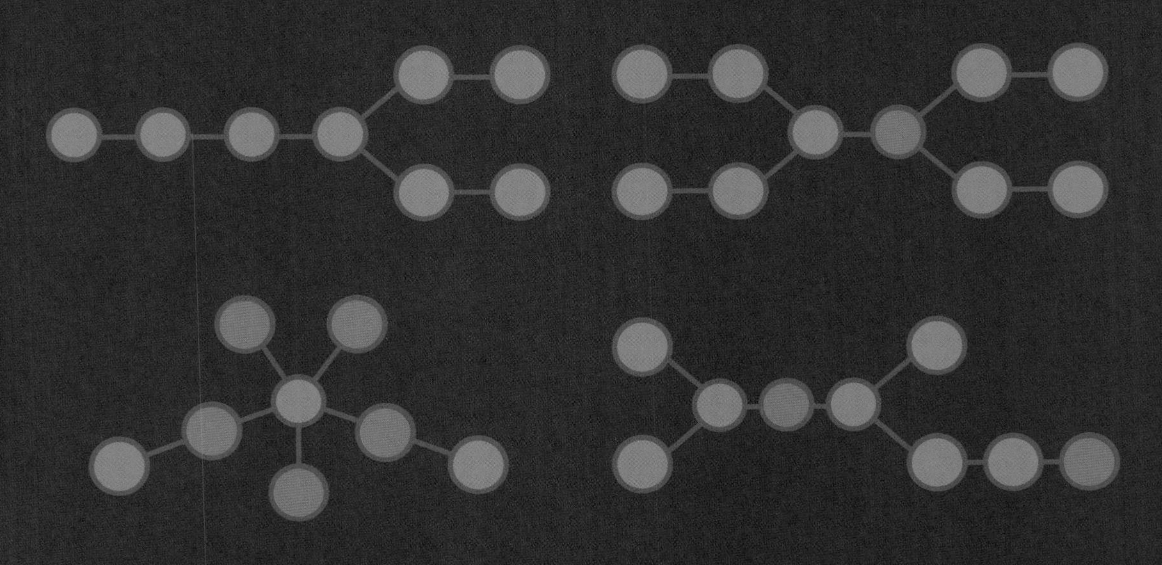

There are two arrangements of eight lily pads that cannot host a party on any lily pad:

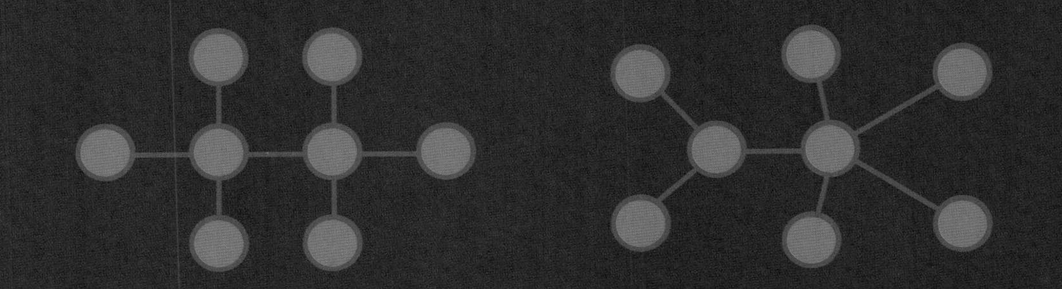

Matej Veselovac is responsible for this delicious extension of the linear puzzle to trees. **Thank you, Matej.** 🙂

I'll leave it to the curious reader to ask questions about:

- where a queen frog can start and still have a successful party.
- graphs other than trees.

COLORING UKRAINE

This is a map of Ukraine.

Regions that share a border are colored differently.

This is the same map of Ukraine. The regions are represented by circles. Any two regions that shared an edge on the previous map are now connected.

Yellow is the most common color.
Pink is the least.

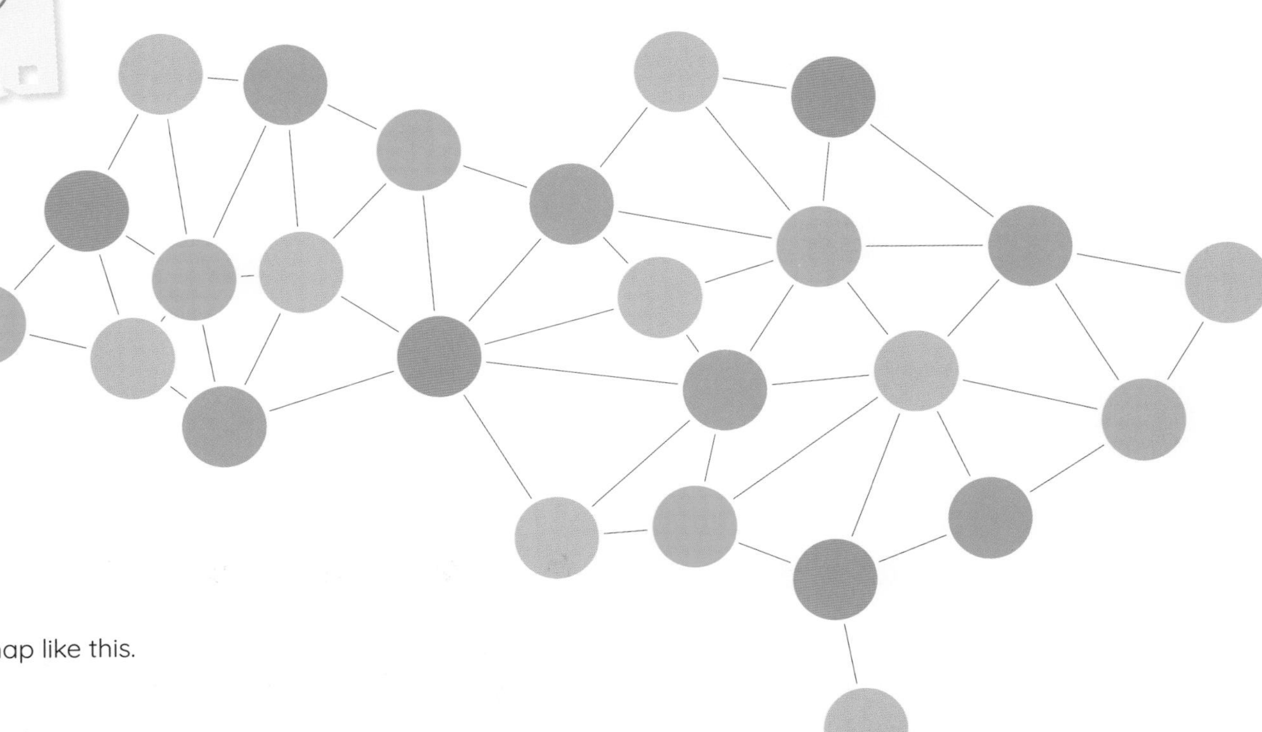

Here are a few steps to color a map like this. We'll call this our recipe:

1. Start with uncolored circular regions.
2. Choose a color. Start painting circles with this color. Make sure no two circles with the same color are connected. Repeat until it is impossible to fill any more circles.
3. Repeat "2" until the map is totally colored.

1.

Starting with the color **blue,** we follow the recipe.
Can we stop using **blue** now?

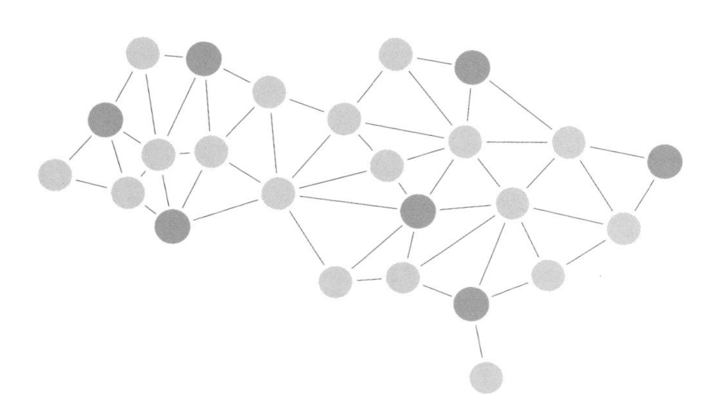

2.

No, we must continue painting with **blue** until there are no new circles that can be painted **blue.**

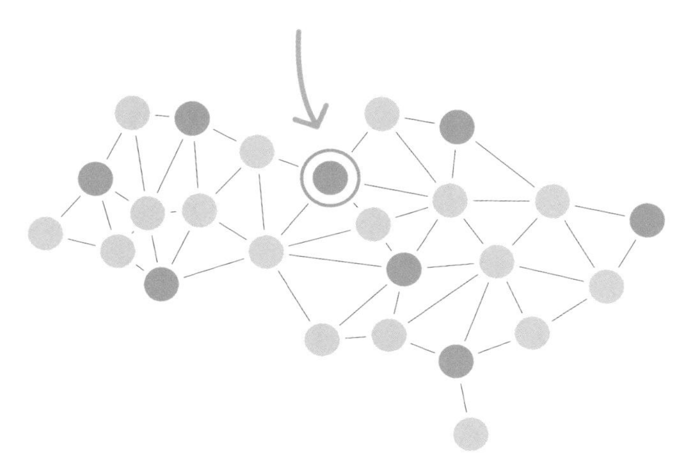

3.

We painted 8 circles **blue.** Let's put that "8" in a prominent position and start painting with a different color.
Let's choose **yellow.**

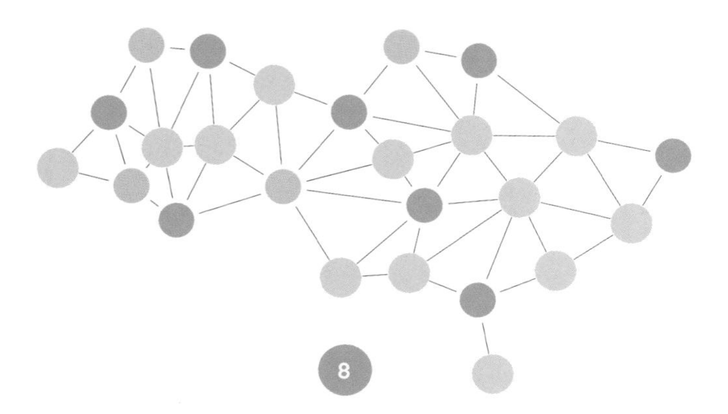

4.

We kept painting until we got seven **yellow** circles.
Can we paint another **yellow?** No—any additional circle we painted would be connected to another yellow circle.

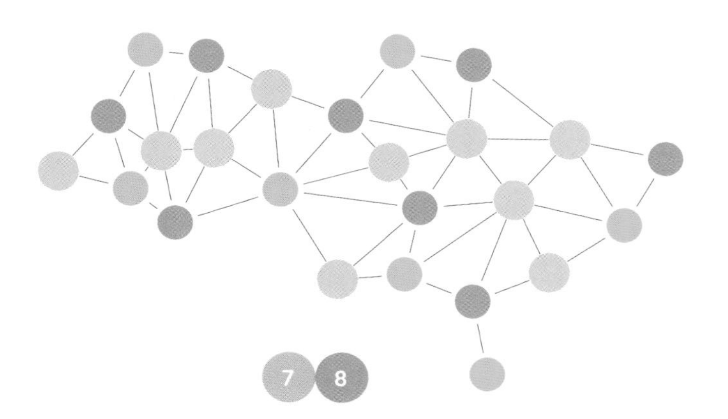

5.

Next we color 6 light green circles. You notice we are building a number as we are coloring our map. Our number now is 678, but we are not finished yet.

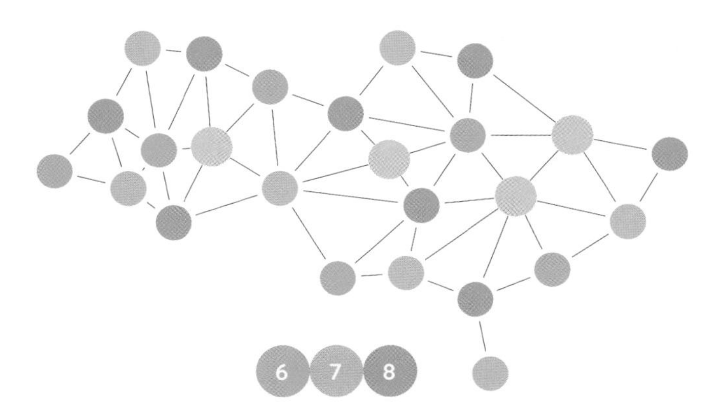

6.

We add three **dark green.** There was actually no way to add more or fewer. Our number is now 3,678.

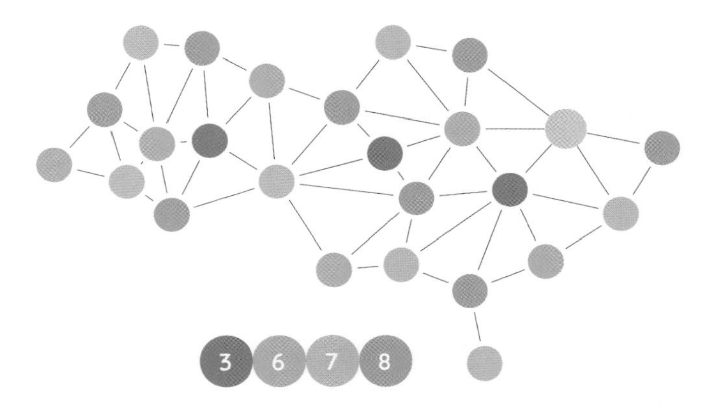

7.

Lastly we add a single **pink.** Our map is finished. It gets the number 13,678.

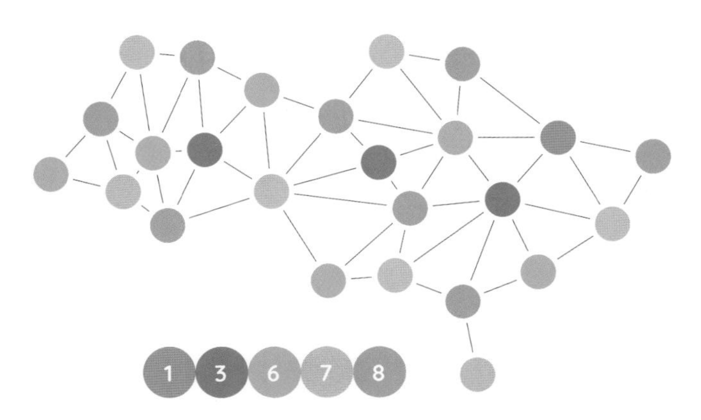

The **blues** are the number of units.
The yellows are the number of tens.
The light greens are the number of hundreds.
The **dark greens** are the number of thousands.
The **pinks** are the number of ten thousands.

- What is the biggest number you can get by coloring Ukraine?
- What is the smallest number you can get?

I'll share my highest and lowest numbers in a couple of pages.

SPOILER ALERT!

8.

Sometimes you will end up with ten or more circles of a color. That's fine. It just means you have a little thinking to do. What is the number for this coloring?

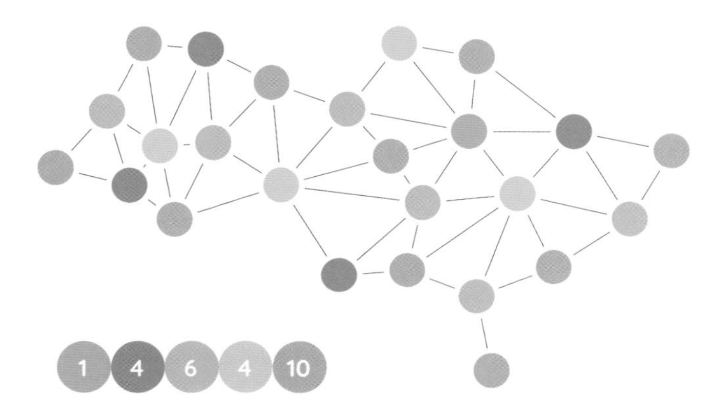

There are 10 units **(blues)**
There are 4 tens (yellow)
There are 6 hundreds (light greens)
There are 4 thousands **(dark green)**
There is 1 ten thousands (pinks)

Total: 14,650

In order to get a big number you actually want to color fewer circles at each step of our recipe! This is a fun paradox for the kids to figure out! Don't spoon feed them, but ask them to reflect if you find them repeatedly stumbling on a small number while they were trying to find a big number.

SPOILER ALERT!

Rather than jump into a complicated map like Ukraine, you may choose to color a simpler map like the ones below.

These simple maps encourage children to make generalizations that are impossible on the more chaotic real-world maps. **Real-world puzzles and problems are not always the best.**

Color river maps of length 2-9. What are the smallest and biggest numbers possible? Fish keep these maps in their pockets in case they get lost on the way to their spawning grounds.

Ducks like to swim circles in lakes with a perimeter of 3-10 circles. Some primates like to walk around 3x3, 4x4 and 5x5 towns and villages like the one on the right.

What are the smallest and biggest numbers possible?

SPOILER ALERT!

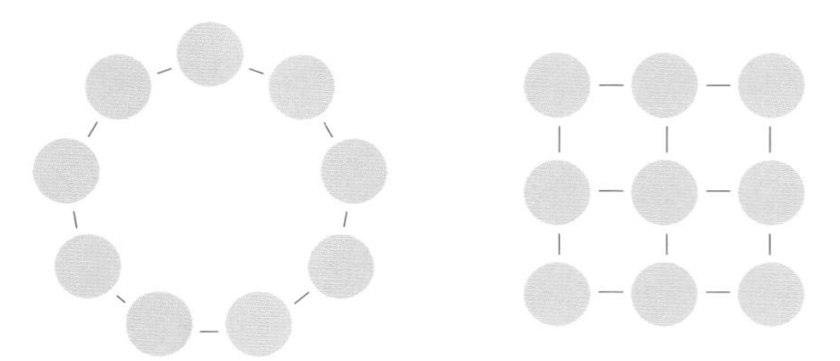

9.

The smallest number I've discovered for Ukraine is 2670 + 10 = 2680.

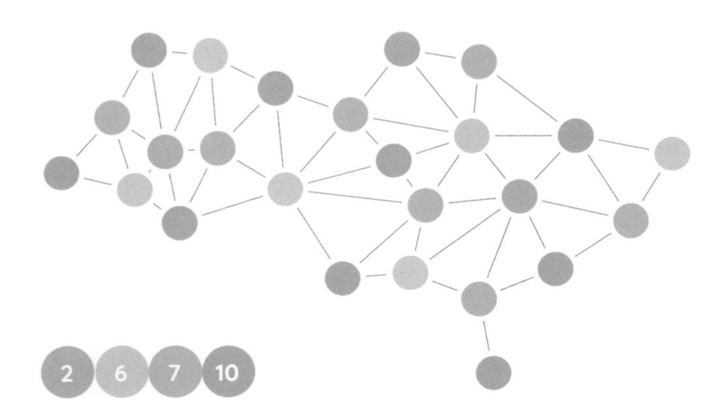

10.

The largest number I've discovered is 144,466.

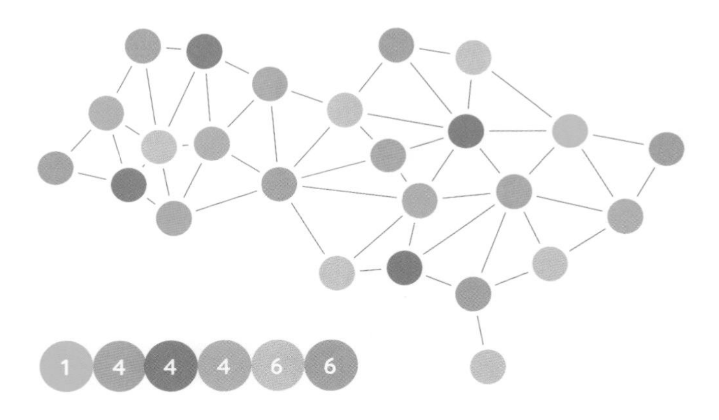

11.

Rivers can be painted with two colors. Here are the lowest numbers possible. See the pattern?

4 5
4 4
3 4
3 3
2 3
2 2
1 2
1 1

12.

All rivers of length 4+ can be painted with three colors. Here are the highest numbers possible. The pattern is a little more complex.

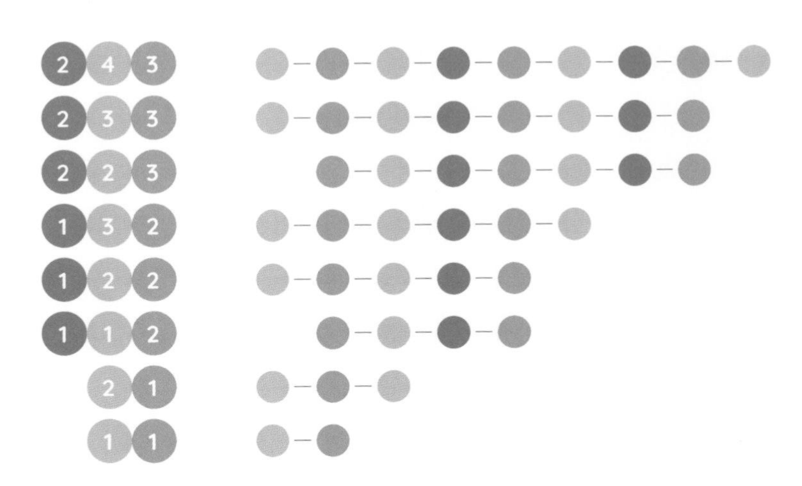

13.

Lakes can be painted with two or three colors depending on whether they have an even or odd perimeter. Here are the smallest and biggest numbers:

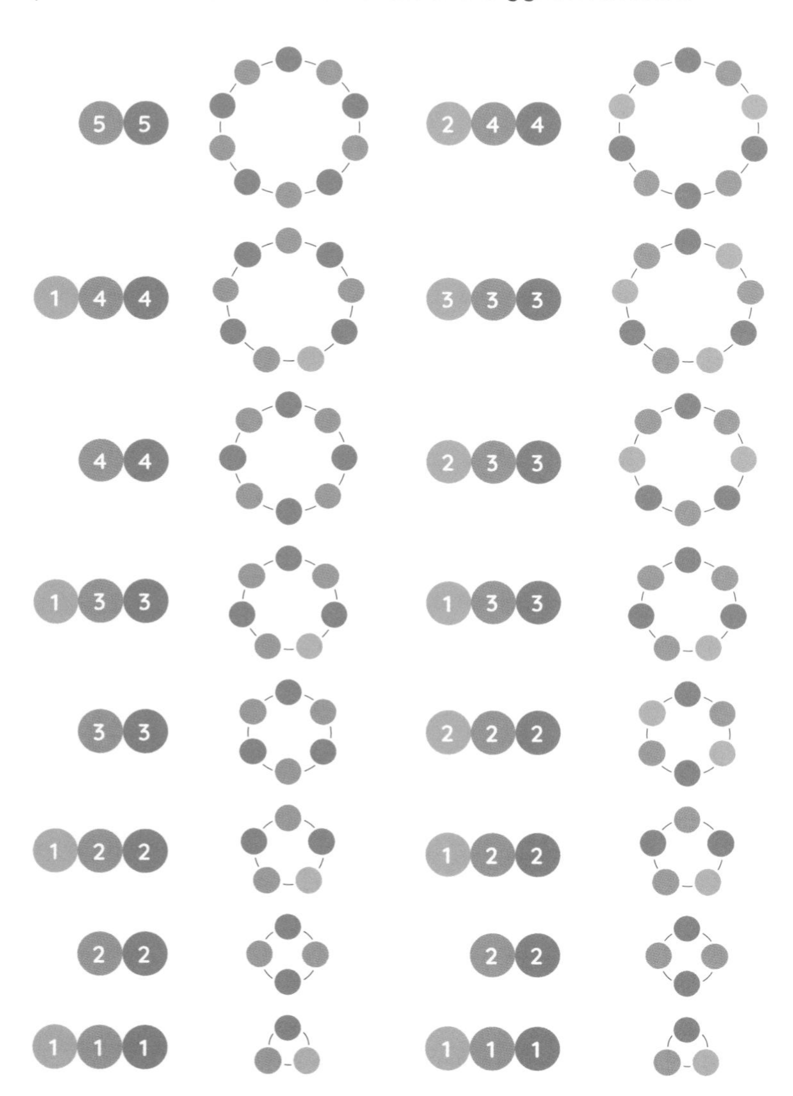

All towns and villages can be colored with two colors. The lowest numbers all have a coloring with a checkerboard pattern.

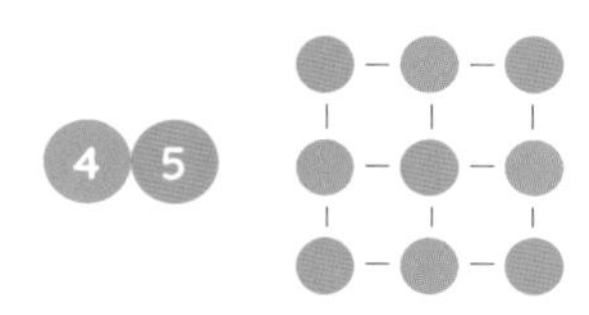

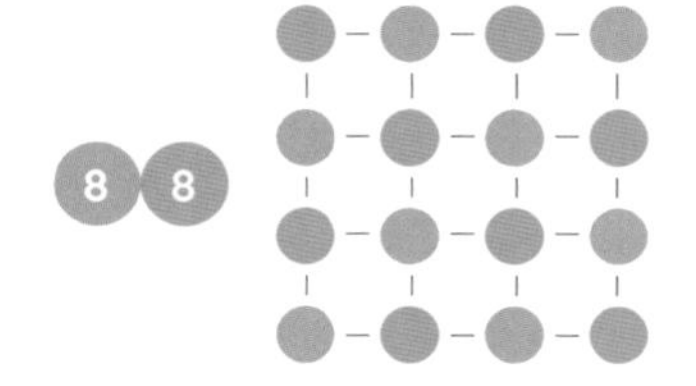

The largest number is much harder to find. I'm not confident that these are the best, but they took an hour to find and that's enough. 🙂

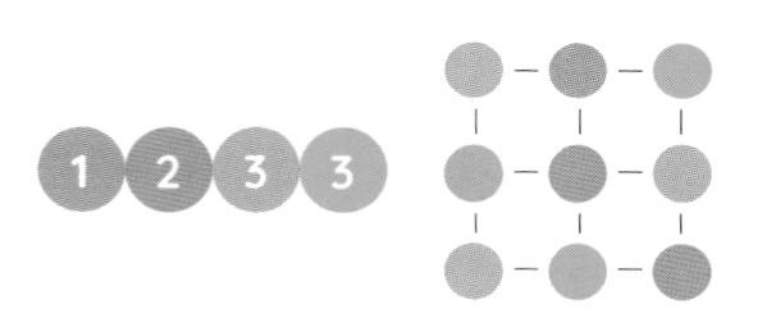

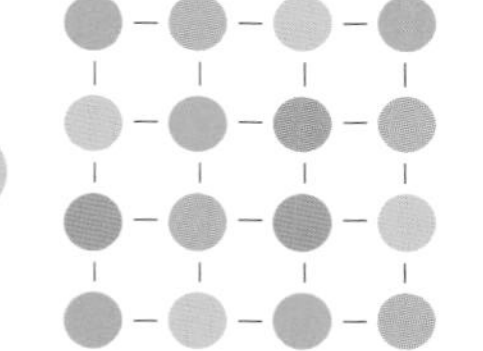

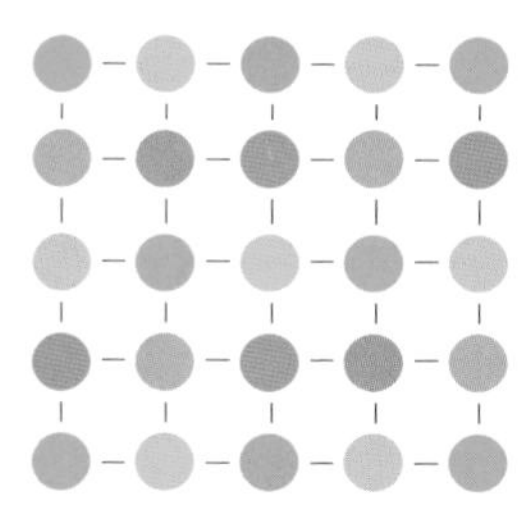

Using as few circles as possible, design a map with no overlapping connecting lines that has its largest number greater than one million.

SPOILER ALERT!

Have beauty as your highest standard. Students should be proud of their work. For many this will mean colorful puzzle-sheets. For others it will mean number patterns that are sharp and precise.

I introduced my daughter Acadia to "Coloring Ukraine." Her first response was "There are that many provinces in Ukraine?" But then she got busy trying to beat Gord! We chatted about where Russia was currently invading and besieging and some history of the country and stories of when I visited Ukraine in 1993. When we were satisfied with coloring Ukraine, I pulled out a map of Guatemala subdivided by its 22 departments. I pointed out where Acadia was born and she once again engaged the challenge of both maximizing color usage and the number of colors. As she hasn't particularly enjoyed math this year, I'm filled with joy when she engages math in more creative and relational ways.

Zaak and his young family lived in Guatemala in the early 2000s where they built schools in a Mayan-speaking area in the central mountains. Acadia was born there.

Zaak and **Acadia** and their whole family now live in Calgary, Canada, where Zaak is a high school teacher and Acadia is a junior high school student.

Here is a map of only ten circles that can be painted with our algorithm in such a way so as to make a number greater than a million. **Do it!**

SPOILER ALERT!

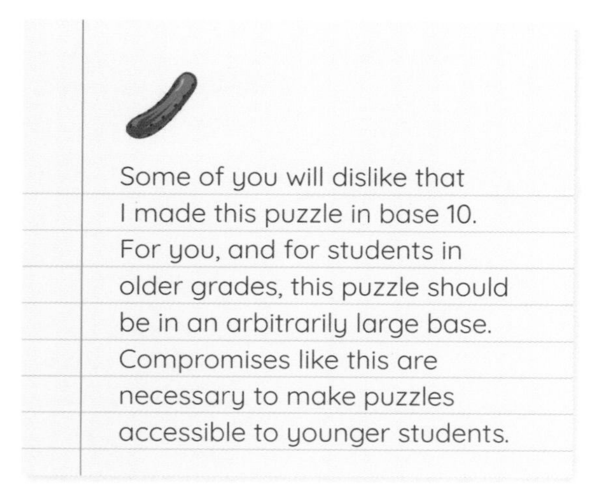

Some of you will dislike that I made this puzzle in base 10. For you, and for students in older grades, this puzzle should be in an arbitrarily large base. Compromises like this are necessary to make puzzles accessible to younger students.

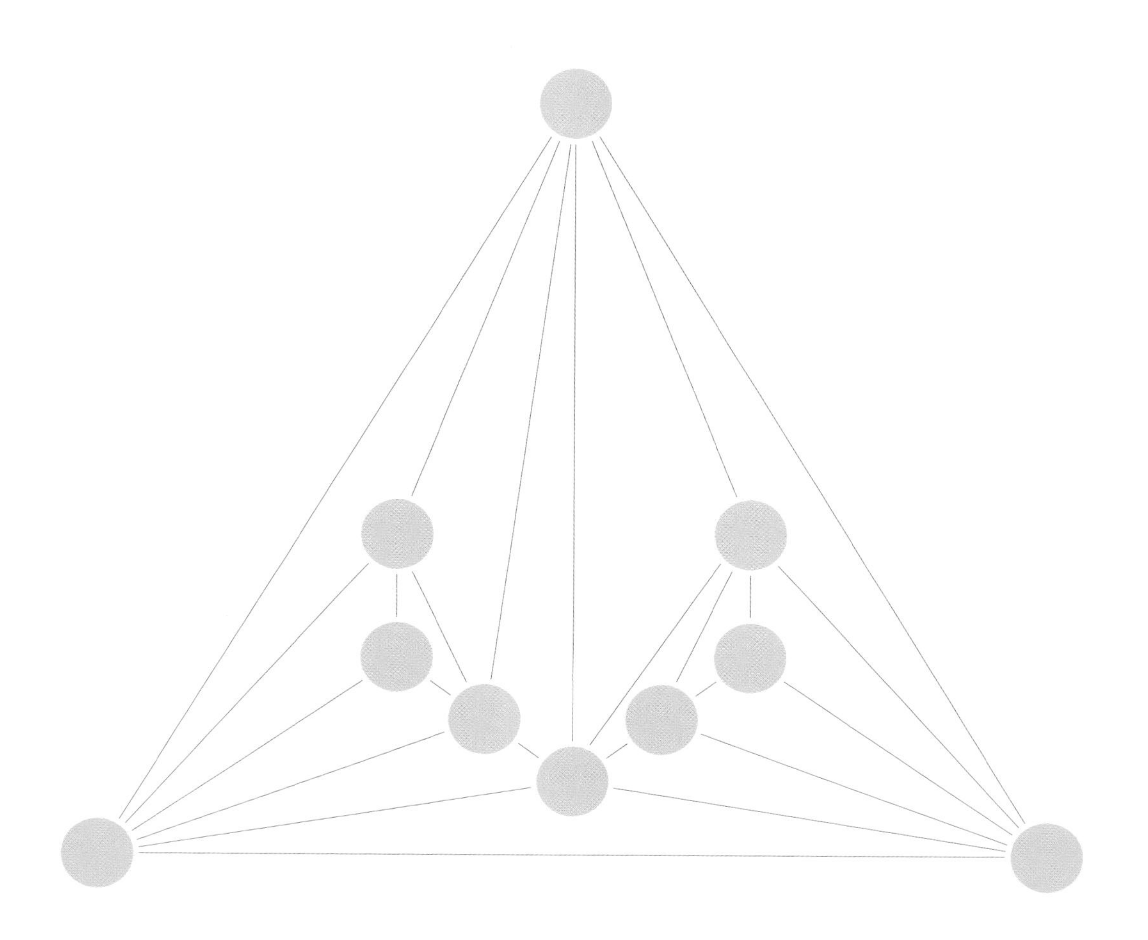

A pair of kids are finished coloring a map, but you are busy with another pair. What can you quickly do to keep everyone engaged? Here are some options:

- Another pair are also finished. You motion the two pairs to compare answers.
- Hand the pair a map of another country.
- Ask them to design their own map or consider simplified "countries" like the rivers and lakes.

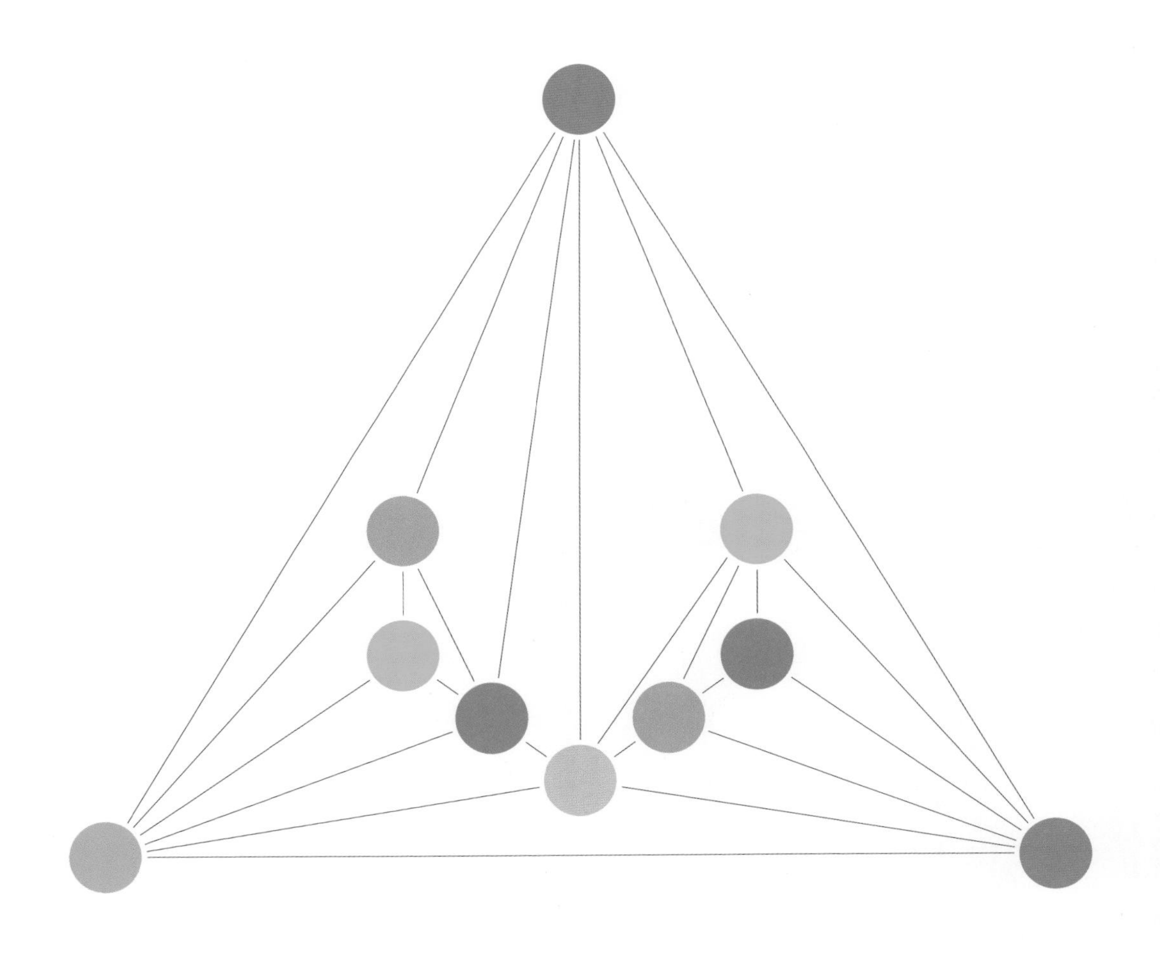

SCALE A MOUNTAIN!

Enough screen time!
It's time to scale a mountain!

Fill in each circle at the base of the mountain with an integer—zero or bigger. You can't use the same integer twice as you're scaling.

The next row up may not all be possible to climb. If possible, fill each circle on the next row with integers that are right in the middle of the two under them.

Right in the middle of 0 and 8 is 4. **Good!**
Right in the middle between 8 and 6 is 7. **Good!**

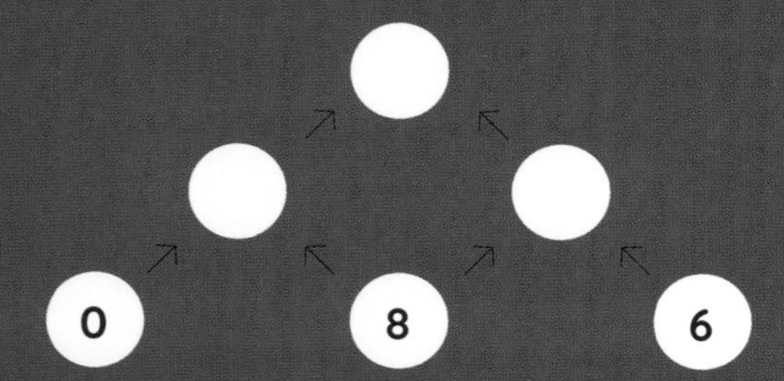

Continue up the mountain. Sadly, the 4 and 7 do not have an integer right in the middle between them… 5 is too close to 4… 6 is too close to 7.

We have failed to scale this mountain.
Find three integers—zero or bigger—that allow us to scale this mountain.

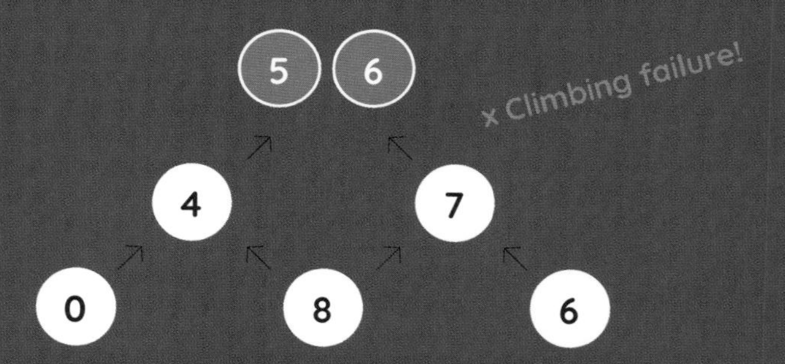

This expedition may look like it's been successful, but something is wrong. It fails because we cannot re-use an integer. **We've used two fours.**

5
4 6
0 8 4

× Climbing failure!

This one works. There are lots of ways to succeed in scaling this mountain.

What is the smallest number you will find at the peak of a successfully scaled mountain? Here we found a 5. Could we have found a 4, 3, or 2?

5
2 8
3 1 15

Find a way to scale this taller mountain. What's the smallest number you will find at each peak?

SPOILER ALERT!

I wish I had been given "Scale a Mountain!" when I was in elementary school because it's just the sort of puzzle I would have enjoyed. It's easy to get started, experiment, explore and make discoveries. And, as a parent with an interest in math education, I love that this puzzle provides an engaging way to develop number sense.

Nancy Blachman San Mateo, USA
Nancy is the founder of the Julia Robinson Mathematics Festival. As a child, she loved playing mathematically beside her mathematician father, Nelson Blachman.

Google "Julia Robinson Mathematics Festival"

One of my favorite puzzles at the moment is called Scale a Mountain! I love the way Gord had this feeling that the peak value would be 10. But to find out that a child breaks this makes us realize how smart our children are.

Erik van Haren Sint Anthonis, Netherlands
Erik is Netherland's math psychologist. He is the director of mathplay.eu and the publisher of puzzles and pedagogic tools that help teachers in the classroom.

For a whole year I thought I had found a pattern that gave the smallest peak value for a mountain of any size. It was a beautiful pattern. Here it is. Imagine hiking up the left side of these mountains...

0, 1, 3, 6, 10... = 0, 0+1, 0+1+2, 0+1+2+3, 0+1+2+3+4...

The pattern on the bottom of these mountains is just as beautiful...

0, 2, 8, 18, 32... = 0, (1)+(1), (2+2)+(2+2), (3+3+3)+(3+3+3), (4+4+4+4)+(4+4+4+4)...

That's all great. It's also wrong.

After a year, a student found a 5-high mountain that can be climbed and has a 9 on the top!

I'll show the student solution on the next page.

SPOILER ALERT!

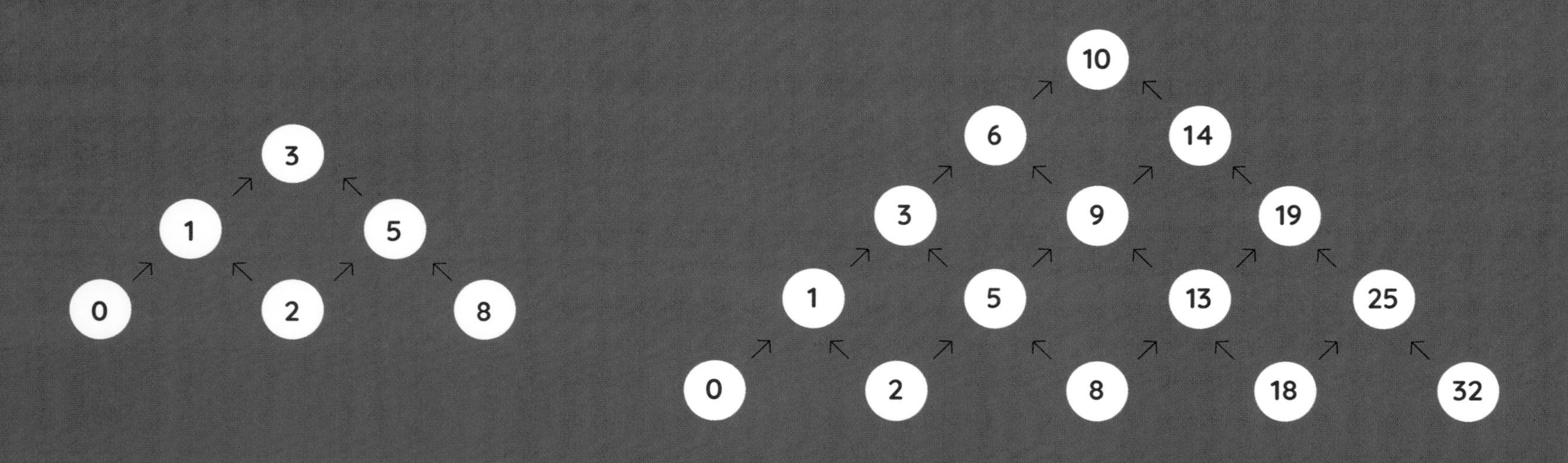

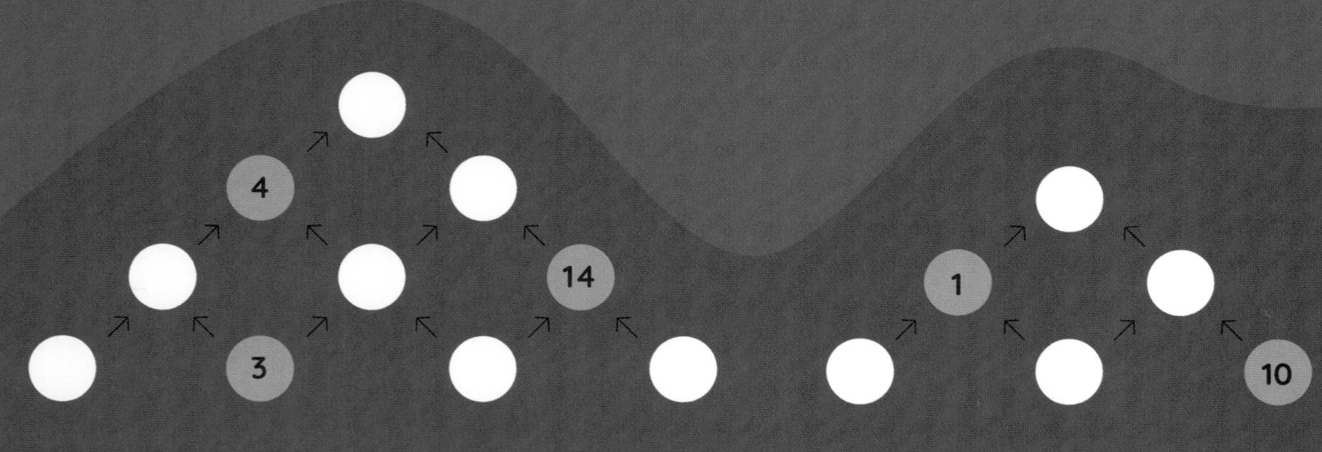

Aaron Holmes—a parent of three elementary school students near Calgary, Alberta, Canada—suggests students will enjoy filling in the missing integers in these successfully scaled mountains.

SPOILER ALERT!

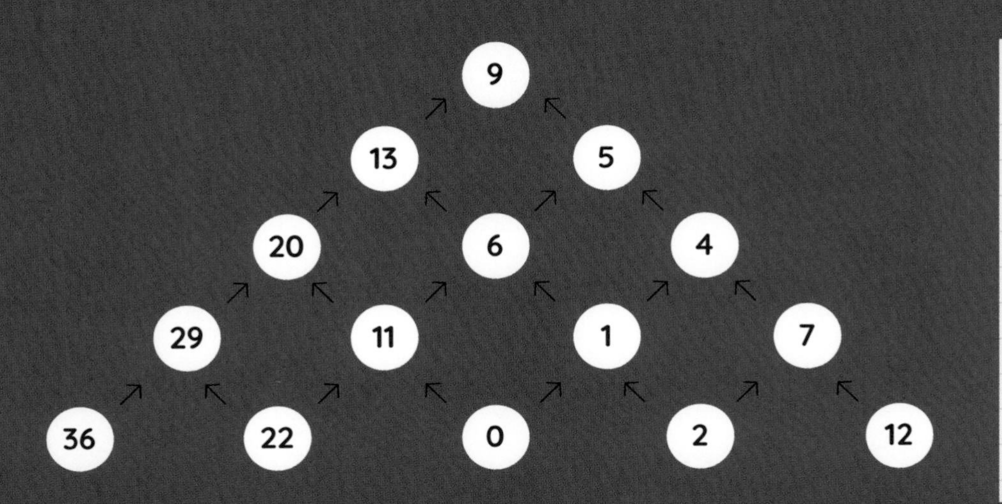

A few years after this student solution, **Brad Ballinger** (Department of Mathematics at Cal Poly Humboldt) found a 5-high mountain with 8 at the top. (Hint: the five integers at the bottom are 0, 2, 10, 24, 56 in some mixed-up arrangement. 🙂

Instead of asking for the peak to be as small as possible, Neil Calkin (School of Mathematical and Statistical Sciences at Clemson University) asks students to find mountains where the largest number is as small as possible.

The students in your class may struggle to climb the 4-high mountain. **That's ok.** They will be able to make it part way up. Do not show them better solutions and burst their bubble. 🙂 Make them think they are the first to climb these mountains! Publicly celebrate each of their mini-successes—especially those of struggling kids.

Students in older grades may enjoy making Aaron-style puzzles. First they will scale a mountain and then erase numbers until just a few remain. **The toughest part is to make sure their puzzle has only one solution.**

Solution to Aaron's puzzles

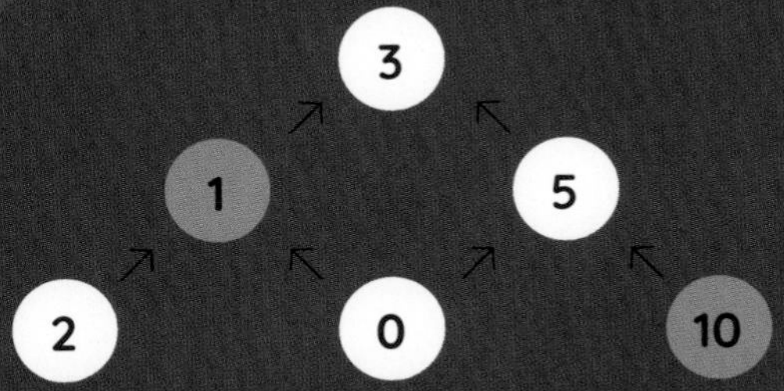

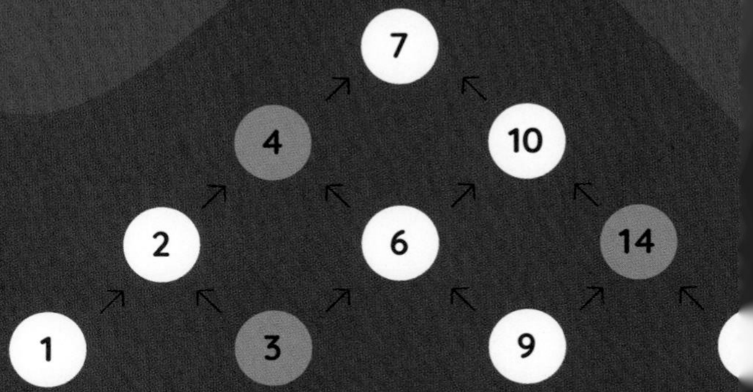

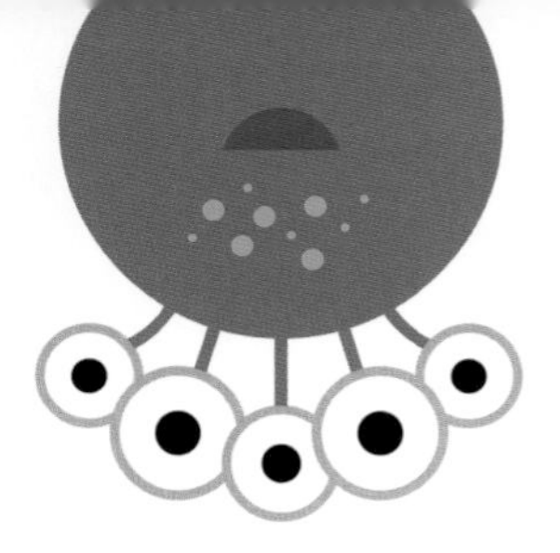

The Greedy Pickle Monster

There are seven jars with 1-7 pickles. Each day, the Greedy Pickle Monster chooses a number and eats exactly that number of pickles from as many jars as possible.

For example, if the Greedy Pickle Monster chooses the number 5, what will happen? He will eat 5 pickles from the jars with 5, 6 and 7 pickles. It won't touch the other jars.

The Greedy Pickle Monster has a lack of discipline. Show that if it thinks, it can eat all the pickles in three days.

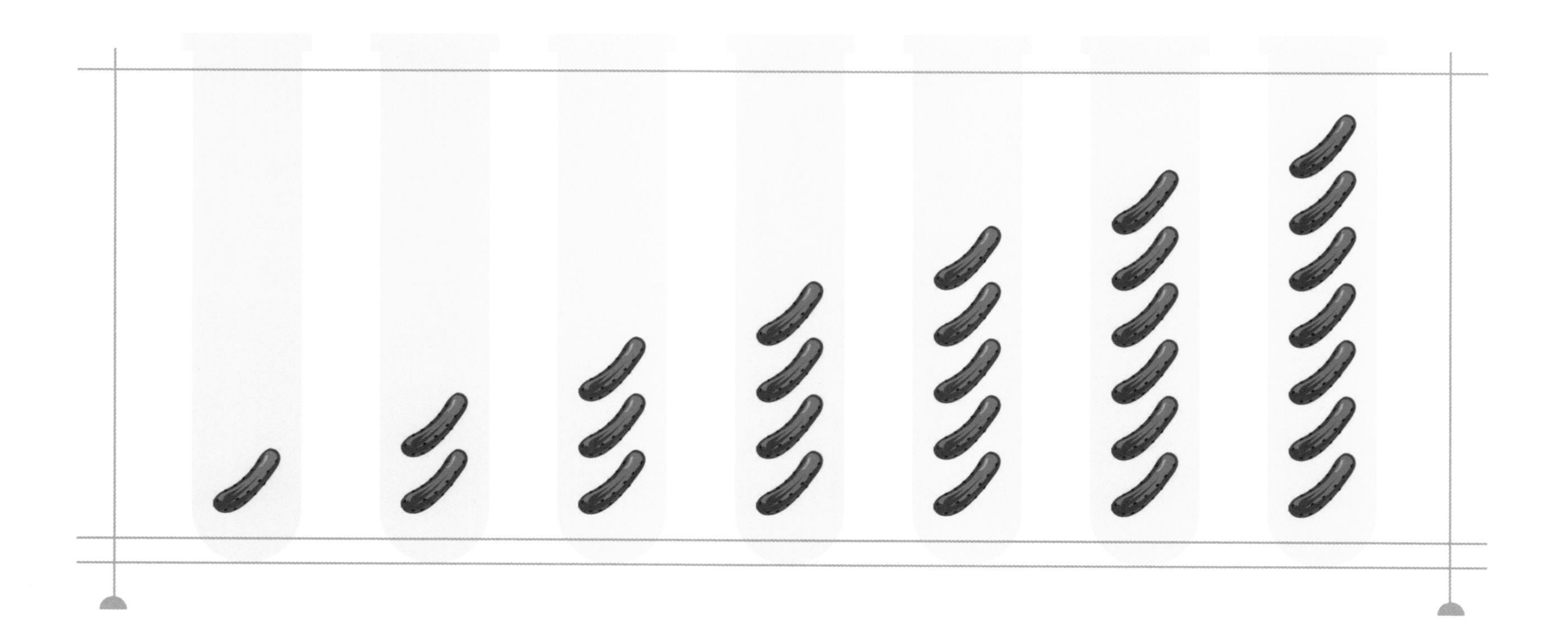

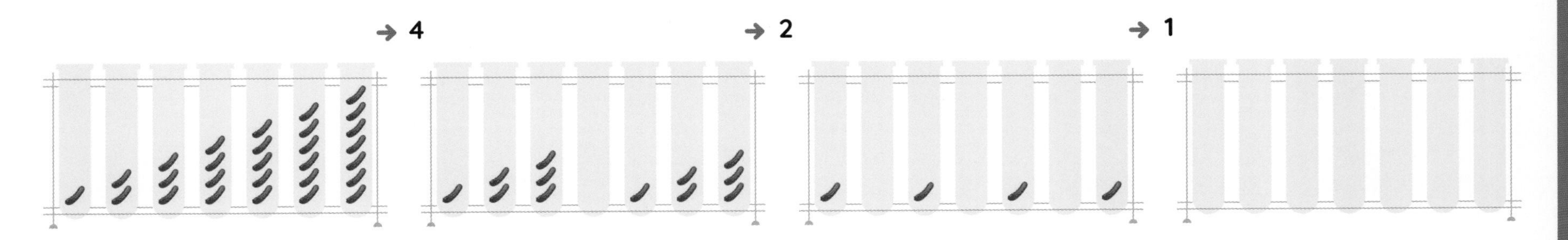

The only way the Greedy Pickle Monster can empty all the jars in three days is first to choose 4, then 2, then 1.

I asked Phillip, a grade 2 boy, to choose a number of pickles to take from these 1-7 jars. His answer was "200 million" with a giggle and a glance around at his peers. **How do you react if a child purposely gives a ridiculous answer?** That child just wants attention.

Your job is just to treat their suggestion as a totally normal answer and move on to the next student: "200 million. That's a lot. Unfortunately you get no pickles, Phillip. Jane - it's your turn. How many pickles would you like to choose?"

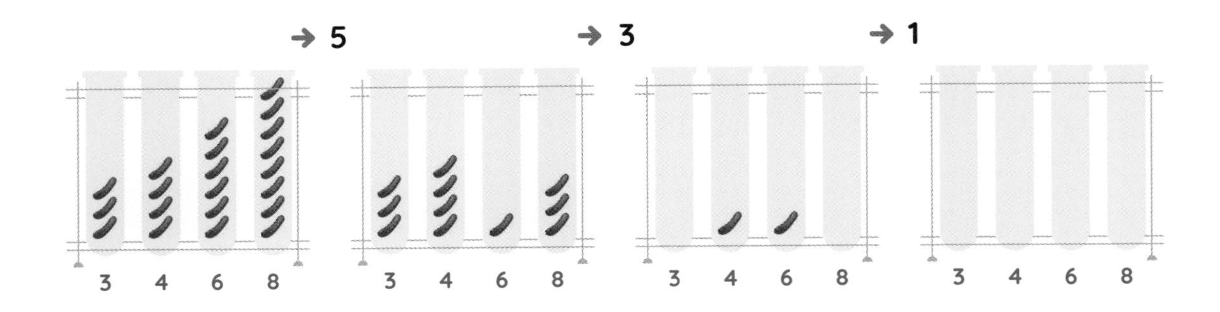

Four jars with at most eight pickles in each can usually be emptied in three days. Four jars with above with 3, 4, 6 and 8 pickles are a perfect example. Find the two examples where the jars cannot be emptied in three days.

SPOILER ALERT!

There are two sets of four jars that hold at most eight pickles and cannot be emptied in three days:

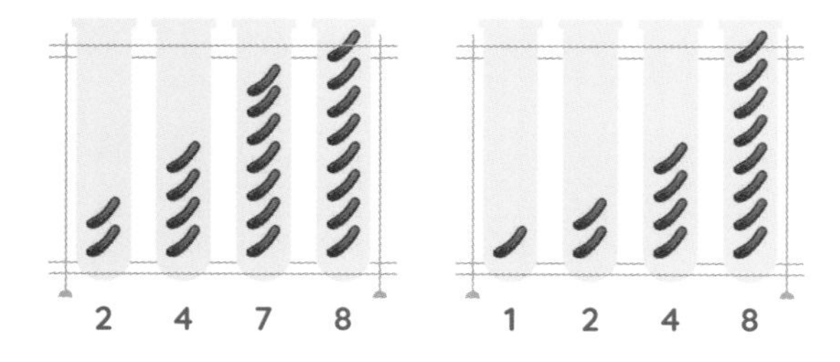

Find a way for the Greedy Pickle Monster to empty these four jars in three days.

SPOILER ALERT!

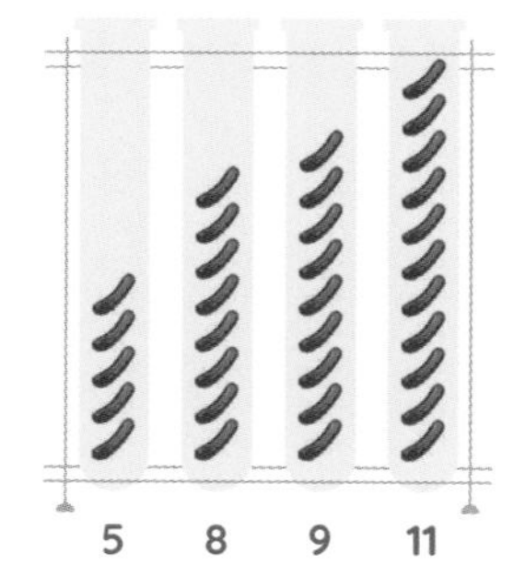

After a while, stop drawing the pickles. I know they're entertaining and you can hear the crunch in your head with every bite, but it becomes cumbersome. 🙂 It took me a long time to find that **{9,11,16,18,19}** can be solved in only four days. **How?**

SPOILER ALERT!

After writing down the numbers in the jars, make sure the students realize that the order of the jars does not matter, but the order that they choose numbers each day matters!

Ask a student to name a number 1-20 of pickles in a jar (or roll a twenty-sided die.) Repeat with other students until you have five jars of pickles. Give your class 71 seconds to find the best way to empty the jars. After 71 seconds: pencils down. Which groups have succeeded in emptying the jars? Which groups have done it in the fewest days?

The 71 seconds is insufficient time for top students to really figure things out so it lets other children shine. **This is time pressure being used in a good way!**

Send these same puzzles home without specifically making it homework. You or some students may find a better answer by the following day. Mentor the behaviour you want your students to emulate.

The first seven prime numbers **{2,3,5,7,11,13,17}** can be emptied in four days. **How?**

The first eleven **{2,3,5,7,11,13,17,19,23,29,31}** can be emptied in five days by many emptying sequences. **How?**

SPOILER ALERT!

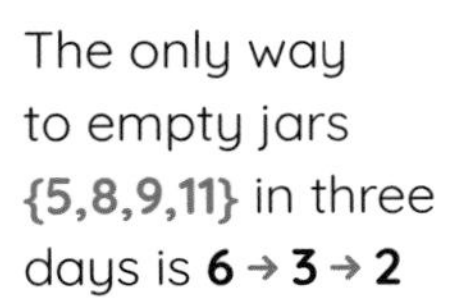

The only way to empty jars **{5,8,9,11}** in three days is **6 → 3 → 2**

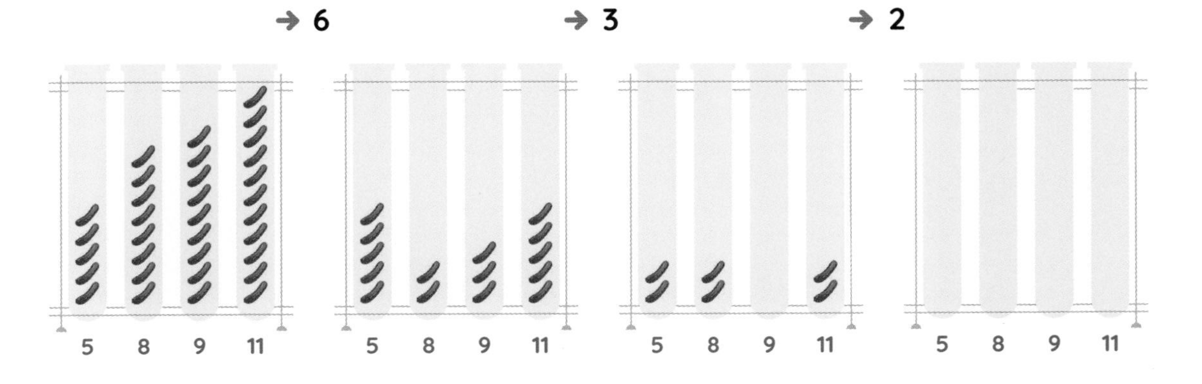

The jars **{9,11,16,18,19}** can be emptied like this: **6 → 10 → 3 → 2**

My elementary school classes follow a pattern. The first fifteen minutes I'm explaining the puzzle. This is not done by me listing the rules. Instead, the focus is on emotional engagement. We usually jump right in with a combination of the class and I failing to solve one instance of the puzzle. Rules emerge gradually.

After fifteen minutes or when I think about 85% of the class knows how to proceed, it is time to create student pairs. How to pair up? Here are some options:

1. **Student-selected-pairings.** (Easy) A danger sign to look out for is social anxiety when unpopular kids are left out.
2. **Puzzle-sheet-pairings.** (Easy) Before pairs are chosen hand out a puzzle sheet to half of the kids in the class. These children then choose a partner to share their puzzle-sheet. Social anxiety is still the primary danger.
3. **Table-pairings.** (Easy) This takes the least time and is the least chaotic. Seating arrangements can be changed-up once in a while to keep things exciting. Watch out for a lack of enthusiasm.
4. **Random-pairings.** (Medium Easy) This technique requires some time each class. For example, teachers have written the names of their students on popsicle sticks which are mixed up and withdrawn in pairs.

The bulk of the class is spent in these pair groups. I wander from group to group being as unhelpful as possible for most, but sometimes offering commiserative tongue-in-cheek whining about how tough the puzzle is. I'm on the lookout for frustration, good ideas and students ready for the next challenge. Quite often we will record discoveries and conjectures on a communal board.

Dan Finkel of MathForLove.com is a math educator I respect enormously. He usually wraps up lessons by bringing the class together for reflection and to talk about their discoveries. I don't do this. In my classes, problems are not packaged at the end. Thoughts are left incomplete and unresolved. This is not because I'm lazy, but because if the class achieves 85% engagement—nothing I can say is sufficiently relevant to a majority of the student pairs to warrant the disruption of their conversations and thoughts. I will interrupt a pair of children, but rarely the whole class. We go to the buzzer!

The first seven prime numbers {2,3,5,7,11,13,17} can be solved in four days . Here are some examples:

3 → 8 → 4 → 2
8 → 3 → 4 → 2
8 → 7 → 3 → 2
11 → 4 → 2 → 1

The first eleven primes {2,3,5,7,11,13,17,19,23,29,31} can be solved in five days in many ways. Here are some examples:

3 → 14 → 8 → 4 → 2
16 → 8 → 4 → 2 → 1
16 → 8 → 5 → 2 → 1
17 → 7 → 4 → 2 → 1
17 → 11 → 4 → 2 → 1

There is no way to empty the jars above in fewer days so let's call the listed sequences "minimal emptying sequences."

When listing these sequences, we are first writing down those that have smaller leading numbers. For the first n primes, find the first minimal emptying sequence listed. Is the first one listed ever the only one listed? What structures are possible for a unique minimal emptying sequence? For example, can the last number be the largest?

Which sequences are a minimal emptying sequence for at least one set of jars? 1,2,3,4,9 takes five days to empty {1,3,6,10,19} but 10,6,3,1 or 6,3,10,1 can empty the same set of jars in just four days. 9,4,3,2,1 can empty jars with 1-19 pickles in five days. That sequence is at least as fast as any other sequence. 16,8,4,2,1 also takes five days.

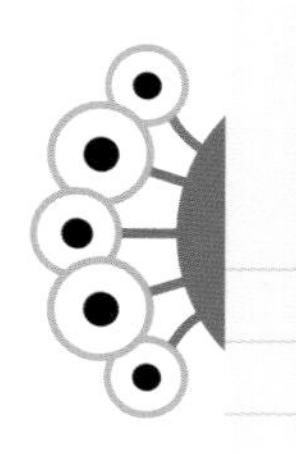

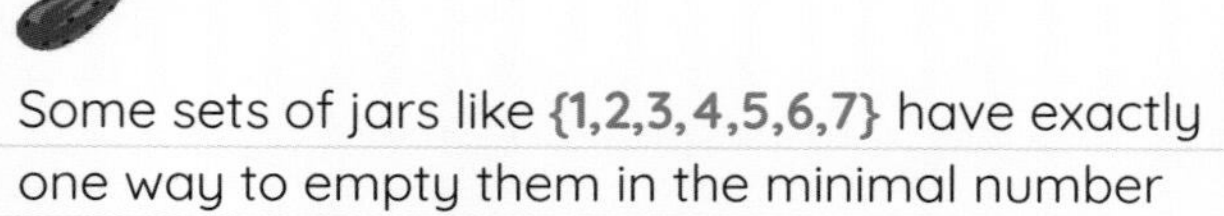

Some sets of jars like {1,2,3,4,5,6,7} have exactly one way to empty them in the minimal number of days. Which other sets have this property?

The original **Cookie Monster® Problem** appeared in the 2002 book *The Inquisitive Problem Solver* by Vaderlind, Guy, and Larson. Richard Guy was my mentor as he approached and surpassed a hundred. He was the oldest practicing mathematician in the world up until one month before his death at 103.

In the original problem, the **Cookie Monster** got to choose a number n and then had the additional choice of which jars to take exactly n cookies from.

With the **Greedy Pickle Monster** version of the problem, we get rid of the second choice. **The Pickle Monster** follows a greedy algorithm meaning that after it chooses a number, it will always grab that exact number of pickles from all possible jars.

Unlike the original, the greedy monster allows us to unambiguously define the operation of an emptying sequence like 6,3,2 on a set of jars.

Richard Guy painted life on a big canvas. He was an avid yachtsman in Singapore, an avid mountaineer in Canada, having a glacial hut named after him and his wife Louise by the Alpine Club of Canada. He still walked into work at age 103. Louise had died ten years before, but on one side of his computer screen was always a picture of her. His book *Unsolved Problems in Number Theory* is the greatest book yet written. 🙂

SQUARE CLIQUES

Squares from the same grade hang around together in the school playground. The 1x1 squares are from grade 1. They will all be connected. The 2x2 squares are from grade 2. They will also be connected. This is true for all grades.

Together, all the squares fill the playground.

The school principal is fair. She demands that these square cliques should all have the same perimeter—or as close to the same perimeter as possible.

For example, if she lets out four sizes of squares into a 10x10 playground—the playground might look like the playground on the right.

Unfortunately, this is not great. The perimeters of the biggest and smallest cliques are quite different.

On pages 316 and 319 I'm going to celebrate some of my failures in trying to solve the 10x10 playground with three cliques. It is easy to fall in love with failures when they are as beautiful as these ;-)

Celebrate failures at every opportunity. You will be rewarded with students who are not afraid to be wrong. Math class will sparkle with conversation.

- The 1x1 squares: perimeter 28.
- The 2x2 squares: perimeter 20.
- The 3x3 squares: perimeter 24.
- The 6x6 square: perimeter 24.

The biggest difference in perimeters is 8. **That's too much.** On this 10x10 playground, find four square cliques where the difference between the biggest and smallest perimeter is as small as possible.

This is the best I could find.
The biggest difference in perimeters is two:

- The 1x1 squares: perimeter 22
- The 2x2 squares: perimeter 24
- The 3x3 squares: perimeter 24
- The 4x4 square: perimeter 24

Perimeter lengths 22/24/24/24 in order starting with the 1x1 region.

I don't think four cliques can all have the same perimeter. For five cliques the smallest difference between the biggest and smallest perimeters in my best solution (below) is six.

Try to find a solution of three cliques on a 10x10 playground so the difference in perimeters is zero or as small as possible.

SPOILER ALERT!

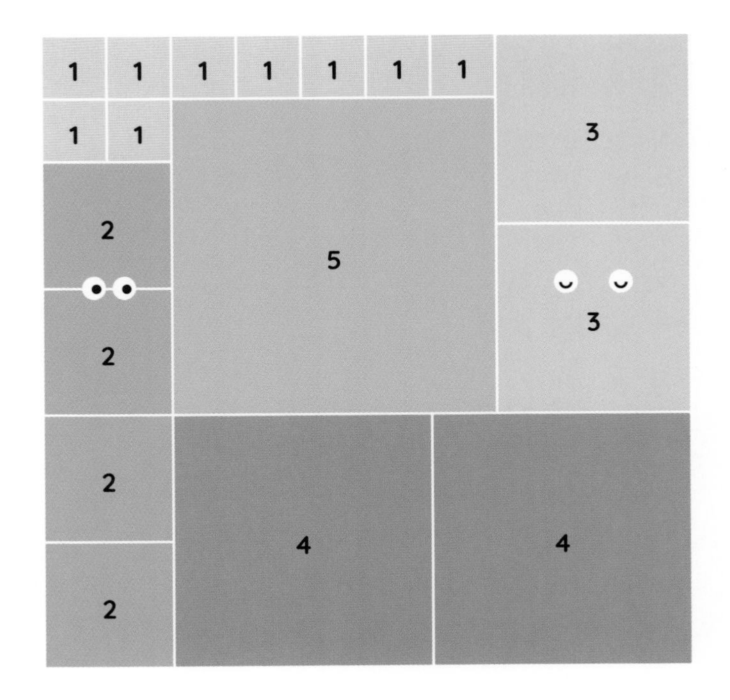

18/20/18/24/20

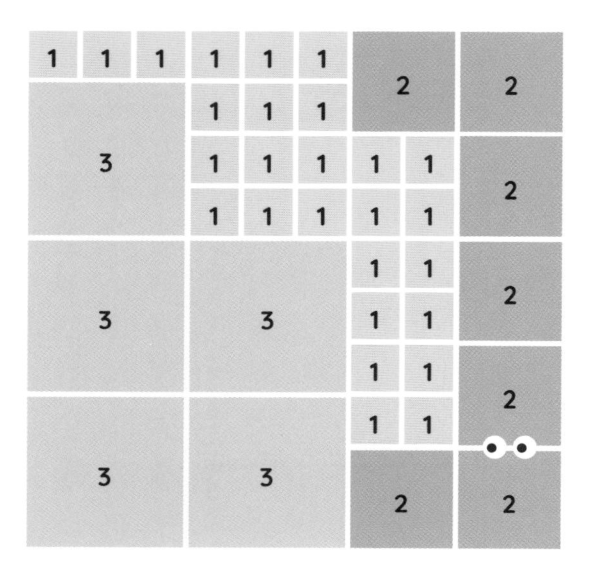

32/32/30

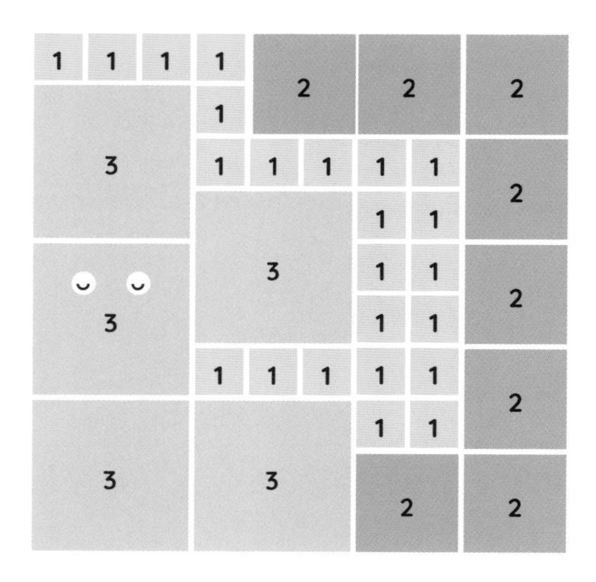

32/32/30

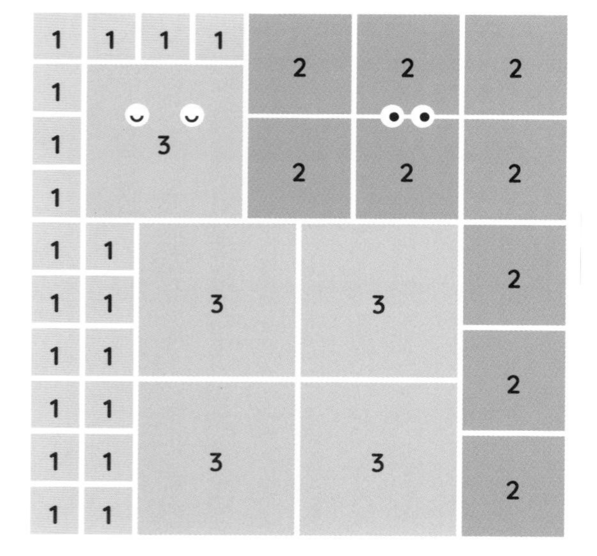

30/32/32

On a 10x10 playground I kept finding solutions where the largest and smallest perimeters differed by two. **These are all failures—it is possible to find a solution where all the perimeters are identical.**

This is also possible for three cliques in an 8x8 and 9x9 playground. Do it. 🙂

SPOILER ALERT!

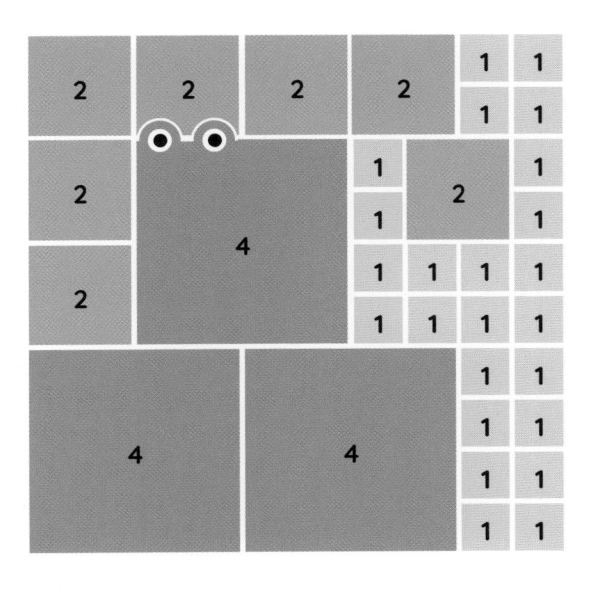

34/34/32

34/32/32

If the perimeters are all equal in size, what are the most square cliques that can play on an nxn playground? **I do not know.**

The number of cliques that can play on a sufficiently large playground is unbounded. For example, a square playground of size n^n x n^n can be covered with n identical rectangular regions of dimensions $n^{(n-1)}$ by n^n. I'll leave you to convince yourself that this is true.

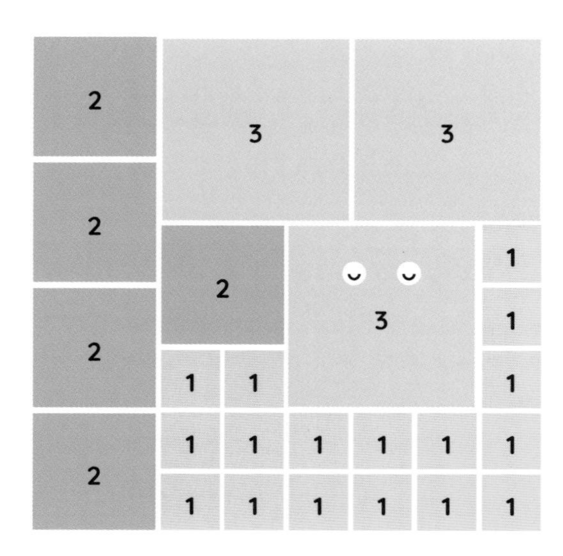

24/24/24

These are solutions for the 8x8, 9x9 and 10x10 playgrounds. All three perimeters are equal on each playground. Almost certainly there are many other solutions.

These are some of the results I am most happy about in the whole book. I am motivated to design puzzles that create beauty and it doesn't get any more beautiful than these. **If you do come up with your own solutions do share them.** 🙂

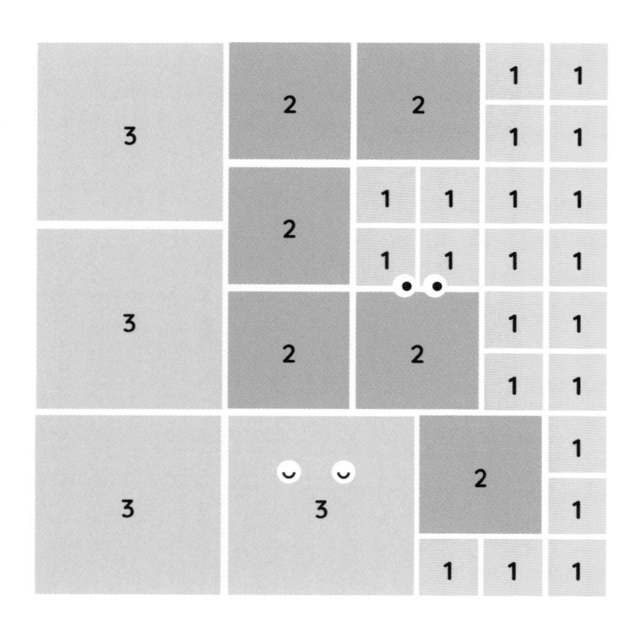

30/30/30

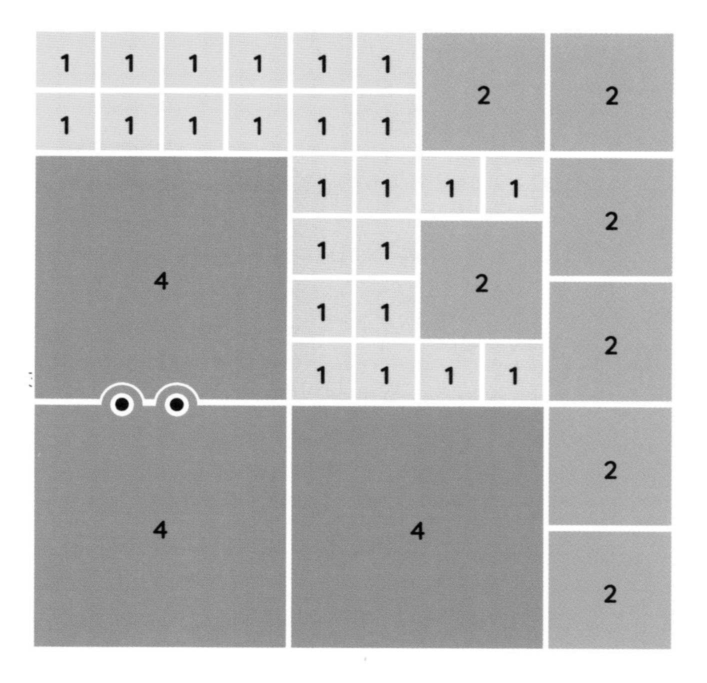

32/32/32

GLUE

Add two squares labeled "1" on a grid.

Each turn you'll be gluing more squares onto the existing ones—starting with a square labeled "2," then one labeled "3," then one labeled "4," etc.

All squares must have their edges lined up on the grid.

No edge length of a square may be greater than the square's label. It may be smaller.

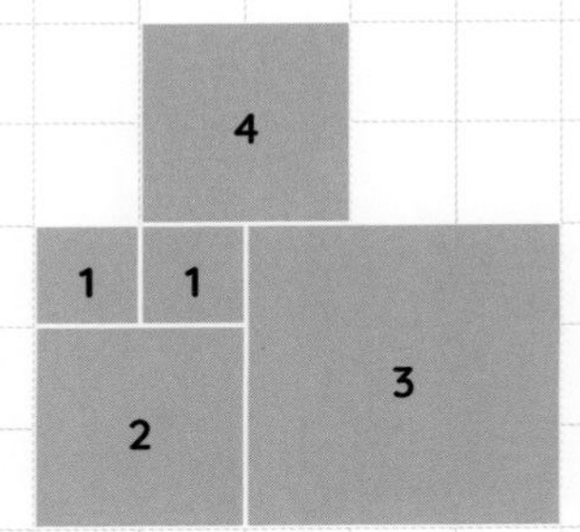

When gluing a square, its label must equal the sum of the labels of all squares sharing an edge. Example: We have just glued the label-4 square. It shares edges with a label-1 and label-3 square. 4=1+3. Good!

How high can you get?

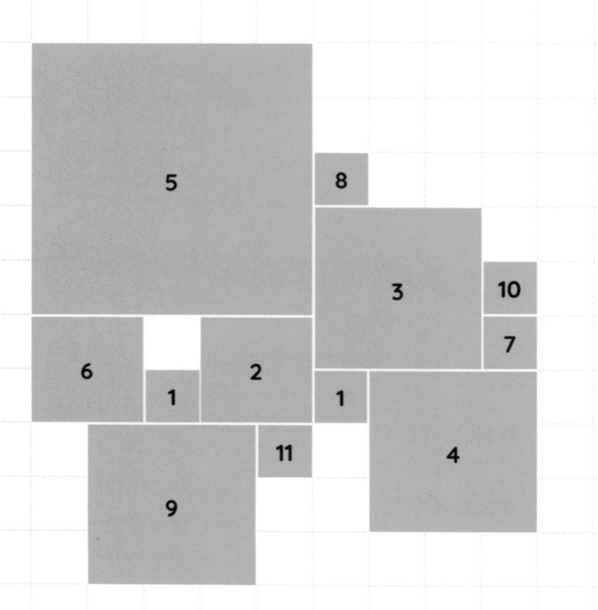

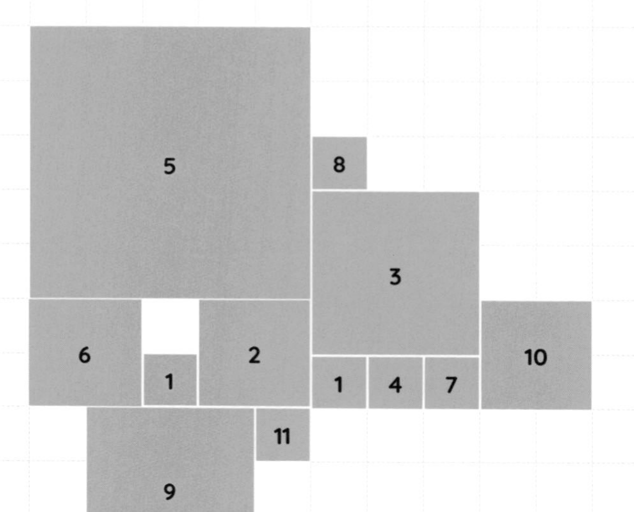

On the left is my first attempt. I failed to get higher than 11. You can see that there is no place to glue the "12" square.

Next to that is my second attempt. I made a small change by shrinking the "4," but that small change allowed me to get to 17. You'll see my solution on the next page.

As usual, there is no need to waste time introducing the problem by explaining the rules. **That's boring!**

Instead, ask a specific student to place two squares labeled "1," then ask the next child to glue a square labeled "2." If they fail to glue it correctly just say that they failed and explain the rule that made them fail. Then ask the next child. Failing in this way is always laughable because they don't know the rules so of course they fail. 🙂

Continue until a child cannot answer because it is impossible. It is important that you still ask the question and keep a straight face even though it is impossible. You can pretend to be ignorant and even ask another child to help the first child solve the problem—implying by your tone of voice that it is possible. In this way you break the child's dependance on you.

Too often children will respond to you based on social clues instead of thinking: Subvert the accuracy of social clues. When a class is particularly good at following social clues, I'll purposely lead them astray with facial and voice clues and then bleat like a sheep at them. **Bahahaha!**

Last page we saw the best way to start explaining a puzzle—just get the students to jump right in before they know the rules. Another option is for you to construct your own example. Which of these two examples is better? There is a correct answer. **Stop and think.** 🙂

The attempt on the top was my second attempt to solve the puzzle. I got up to 17. Your class might be able to get better, but it's not guaranteed. That's the first clue that the example on the top is not the best example to introduce the problem. For motivational reasons, you want average kids to be able to beat your example within their first few attempts.

The second reason that the puzzle on the bottom is the correct answer is that it contains mistakes. These can be uncovered by the students to reinforce the rules of the puzzle. Rules always need reinforcing—this is a great excuse to do it. Get your students to find the mistake where you have placed two numbers instead of just one (hint: the 11s). Get your students to find an addition error (hint: 5+6 = 12?).

The third reason that the puzzle on the bottom is the correct answer also has to do with the mistakes. Presenting a "solution" with mistakes helps remove the stigma of failure from your classroom. The work on the right is a failure. Big deal. We can try again. **That's a nice segue to pairing off your students and letting them try to do better than you.** 🙂

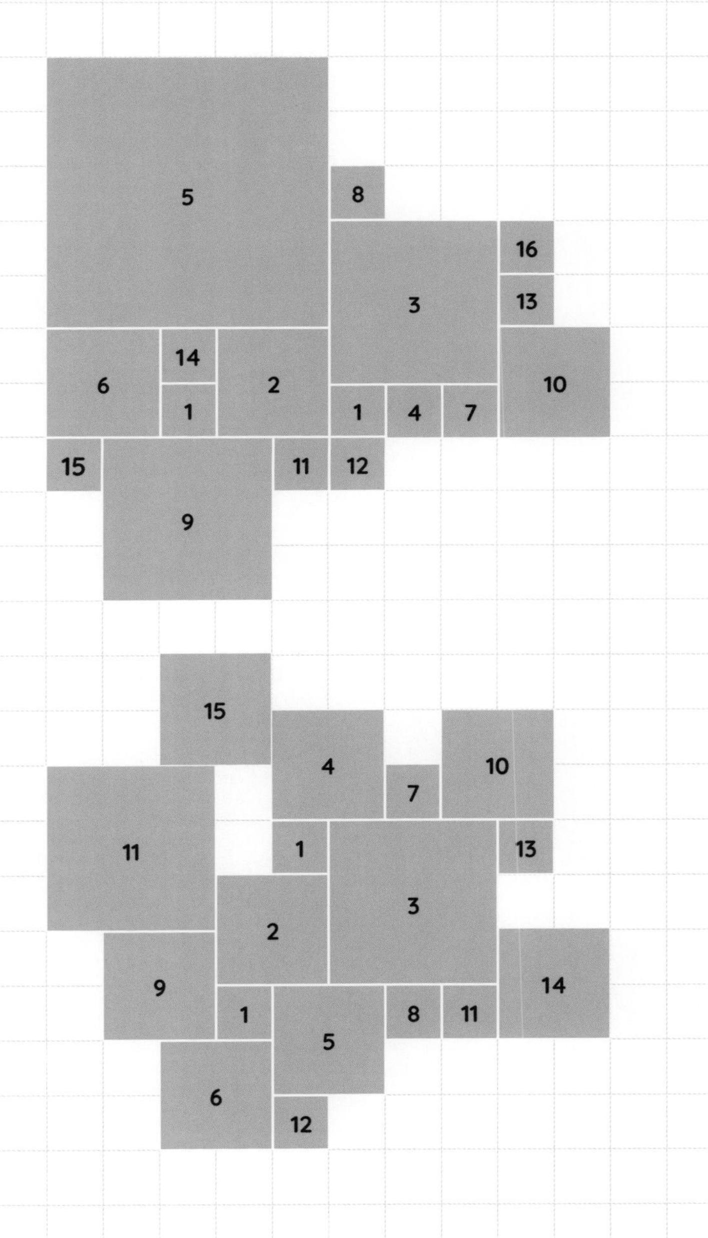

Grade five students Nicholas, Ryan and Nathan beating the previous record by getting to 29!

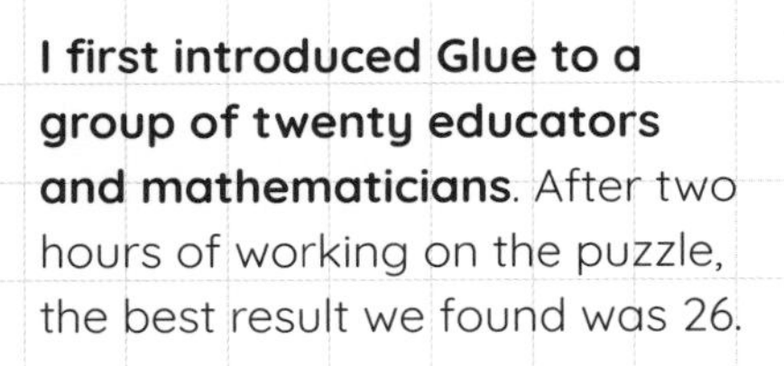

I first introduced Glue to a group of twenty educators and mathematicians. After two hours of working on the puzzle, the best result we found was 26.

I was floored when eight months later a group of three grade-five students found this solution with a high of 29!

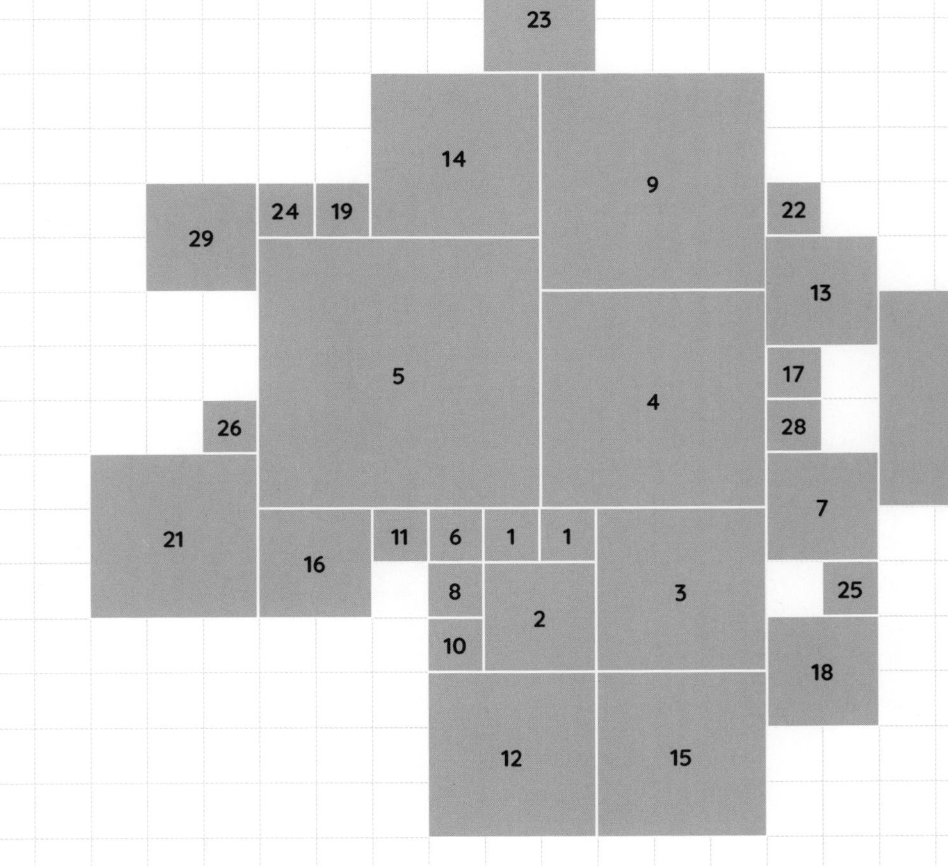

The best solution now is held by thirteen-year-old Tylan of Innisfail, Alberta, Canada. **Instead of showing the solution I'll leave it for you to reconstruct based on the following clues!**

2=1+1
3=1+2
4=1+3
5=1+4
6=1+5
7=3+4
8=2+6
9=4+5
10=3+7
11=4+7
12=2+10
13=5+8
14=5+9
15=4+11
16=7+9
17=7+10
18=5+13
19=5+14
20=8+12
21=8+13
22=10+12
23=7+16
24=5+19
25=9+16
26=10+16
27=10+17
28=9+19
29=5+24
30=7+23
31=13+18
32=5+6+8+13
33=9+24
34=12+22

This is not yet an infinite pickle. It is a single puzzle. That comes with risks. In front of their class a grade 5 boy asked "What's the best answer so far?"

My first inclination was to decline to answer because the best result is so good that none of the students could be expected to get half that good within a single period. That would be demotivating.

My second inclination was to lie—giving the answer of 15, which is good, but maybe not too good. I went with this, but the lie was not sufficient to make the puzzle pedagogically great. This became obvious near the end of the class when some student groups had come nowhere close to the score of 15… and other student groups had surpassed it.

Struggling students had their noses rubbed in it.

We can make Glue into an infinite pickle. Instead of always starting with two 1x1 squares, let's try something more general.

- Choose an integer n that is 2 or more.
- Glue n 1x1 squares onto the grid. Label each of them with a 1.
- Glue a square labeled n, then n+1, then n+2 etc. (All previous rules apply.)
- Try to get as high as you can.

Here are my first and second attempts for n=3.
Notice that we do not use a "2" labeled square.

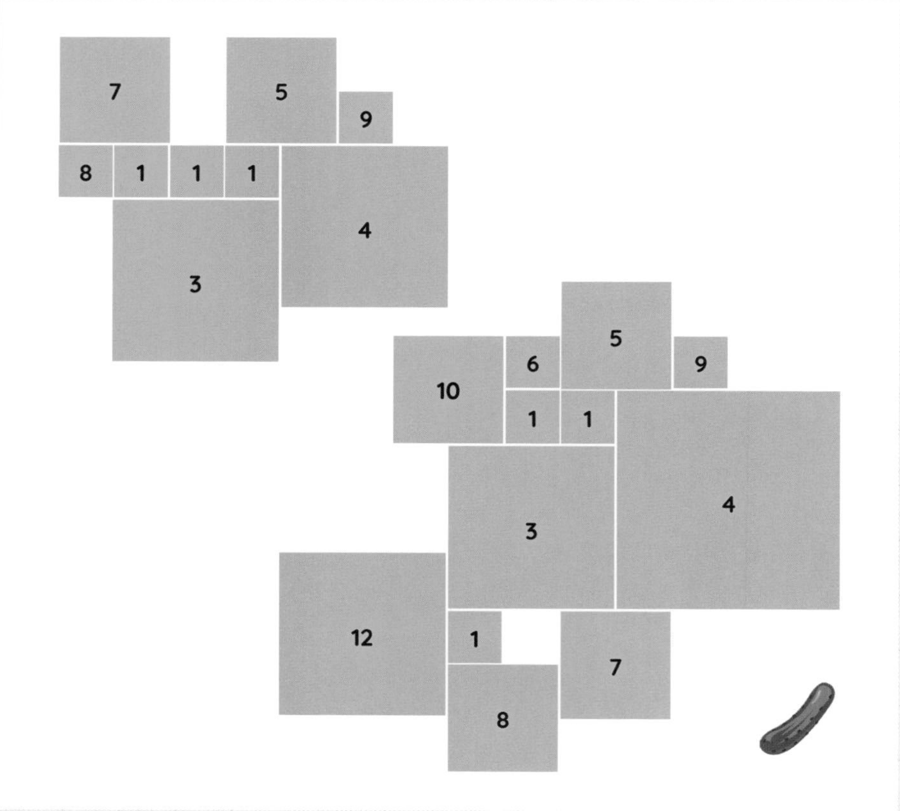

I've been enjoying the Glue puzzle for several weeks, 10-20 minutes at a time, alone or with my husband. "Glue" is a meditative, just-so activity. It gently redirects my attempts to turn it into a hard challenge, taking me back to relaxed math doodling. My initial conjectures proved false; my first attempts at throwing other math at solving the entire thing fizzled. I am enjoying not knowing where this is going.

Maria Droujkova is a parent, curriculum developer and mathematics education consultant. She was born in the Crimea and now lives on the East Coast of the USA.

Grade 2 children are capable of playing with this math, but usually do not have the penmanship to make it beautiful. One option is to make this a 9-minute activity in front of the whole class every math class for a week. Hang up their new creation each day so that by the end of the week they have five attempts to compare.

This problem is deep enough that you could do it for a month and every second would be time well spent.

EARTH ATTACK

Stop the enemy starships from stealing all the Earth's pickles! To shoot one down, all its disk deflectors must be destroyed. You destroy these by hitting them with your ion cannon at a non-negative number of seconds.

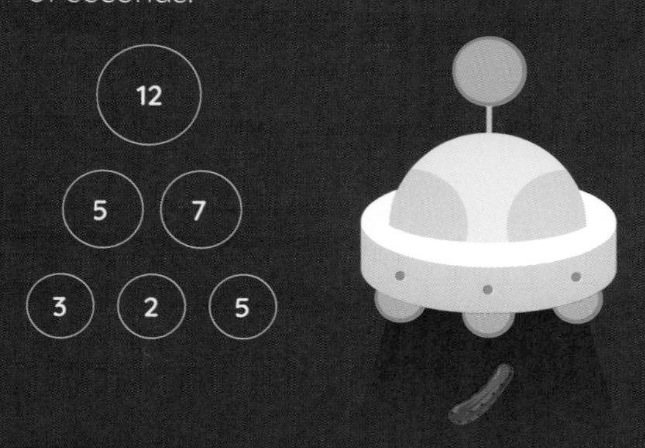

Here you have fired your ion cannon at 2, 3 and 5 seconds, hitting the rear of the attacking starship. Instead of drawing the starship, let's just draw the disk deflectors:

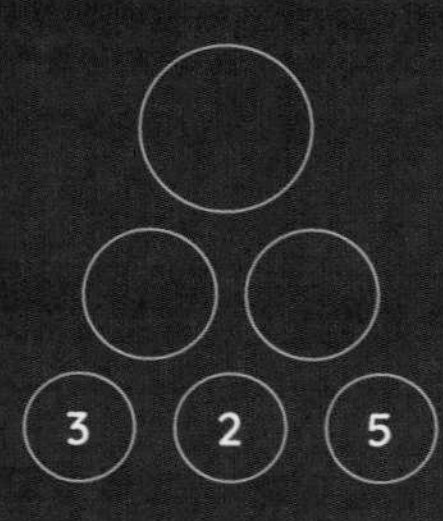

To be successful, you need to hit the other disk deflectors so that they sum to the two disks they are on top of. **Let's complete our attack...**

The 7 looks good because 2+5 =7. Unfortunately for planet Earth, we have only one ion cannon and at 5 seconds you are trying to shoot at two different disks! That's not good. The starship gets through and all the Earth's pickles are stolen.

This is much better. All the sums work (3+2=5 and 2+6=8 and 5+8=13) and there are no duplicates. You've managed to destroy the attacking starship in 13 seconds. The Americas, Asia and Australia have had all their pickles stolen, but the rest of humanity still has theirs. 13 seconds was too long to save most of the planet's pickles, but you're celebrated as a hero/heroine anyways! Congrats!

Perhaps you can destroy the attacking starship a bit faster? **SPOILER ALERT!**

The children must know that this is tongue-in-cheek humour. If this is not your style, rework the story so that you are guiding these starships to land and deliver much needed aid.

Blurring success and failure is a fun part of this narrative. You congratulate them on a job well done, but of course it would have been better if they could have saved the other half of Earth's pickles.

The fastest way to destroy a single one of these starships is **8 seconds.**

What is the fastest way to destroy these three ships? Again, you cannot have duplicate times and your largest time should be as small as possible.

22 seconds is great. Europe and the Americas have had all their pickles stolen, but the rest of the world's pickles has been saved! Congrats!

Finding the best possible score for each of these puzzles is not easy. I'll try to do better than 22 seconds:

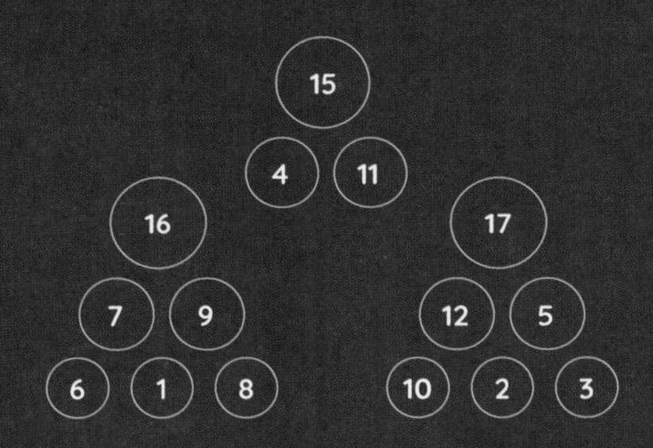

I'd be surprised if 17 seconds is not the best possible, but it's possible I've missed something. I'm also confident that 29 is the best possible for these four starships.

In case the students ask - they can use zero if they wish. They will quickly find that using zero does not help at all. This is for them to discover. Using zero is allowed by the rules.

Bringing a bit of geography into math class is never amiss. **The more variation the better.**

The best time I thought I had found for these 10-disk starships was 56 seconds, but I made a mistake. Find it! **Ed Pegg Jr.** told me he shot the three of them down in 53 seconds.

HOVER-POD ATTACK

Hover pods have an inner and outer ring of disks. The outer ring is created by summing the two closest disks on the inner ring.

This example fails because at both 6 and 7 seconds the ion cannon doesn't know where to aim.

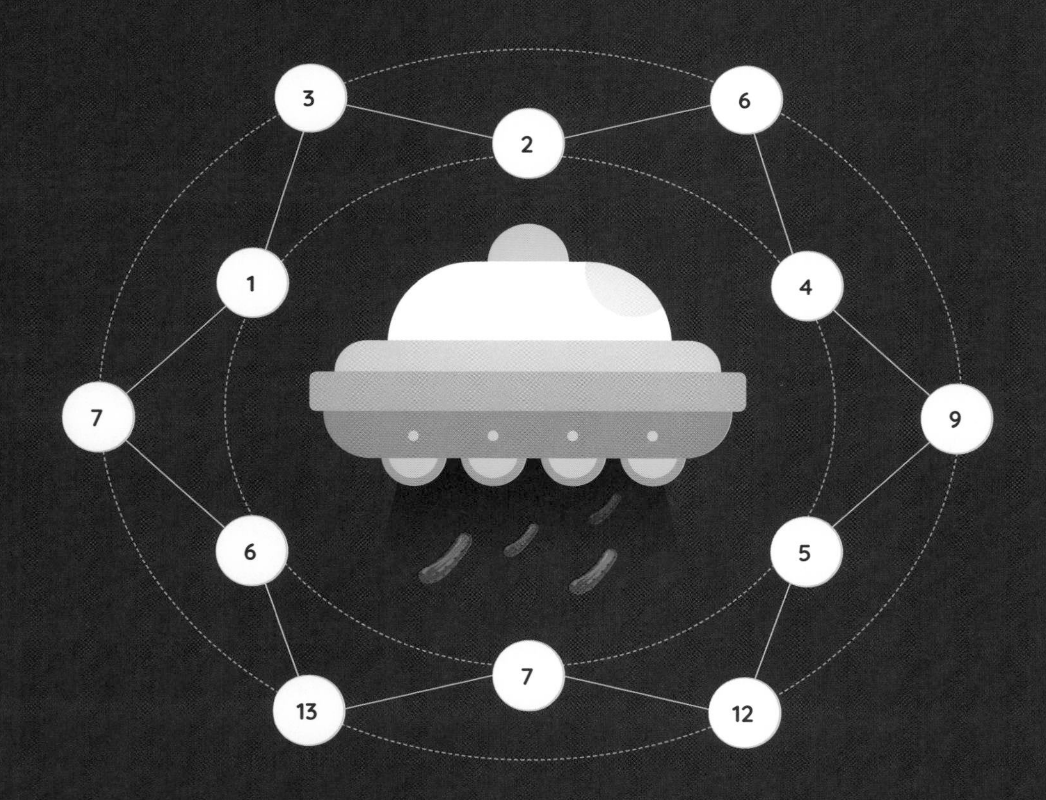

There is room for creativity. What formations do the starships fly in? What does an enemy hover pod look like? Reluctant students can be carried into problem solving by celebrating their designs.

Destroy different sizes of hover pod with the integers 1-6 (below), 1-8, 1-10, and 1-12 (above).

SPOILER ALERT!

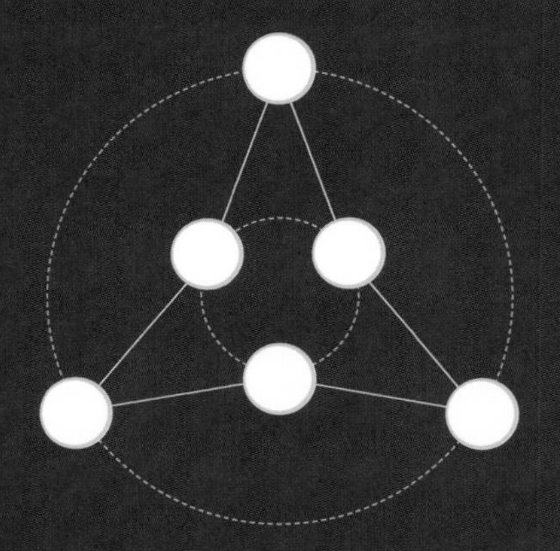

The integers 1-6 can destroy this hover pod with this triangular pattern. The numbers 1-12 can destroy the hover pod on the right with this hexagonal pattern. However, the integers 1-10 cannot destroy a hover pod with a pentagonal star pattern.

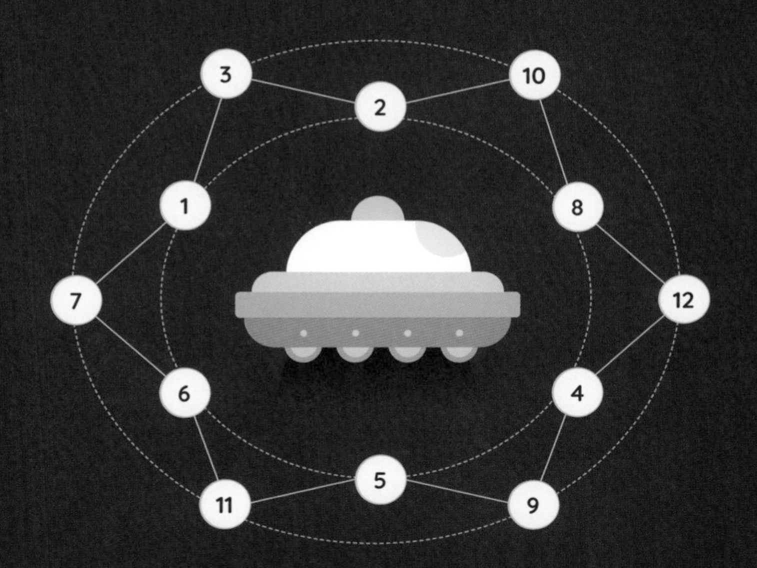

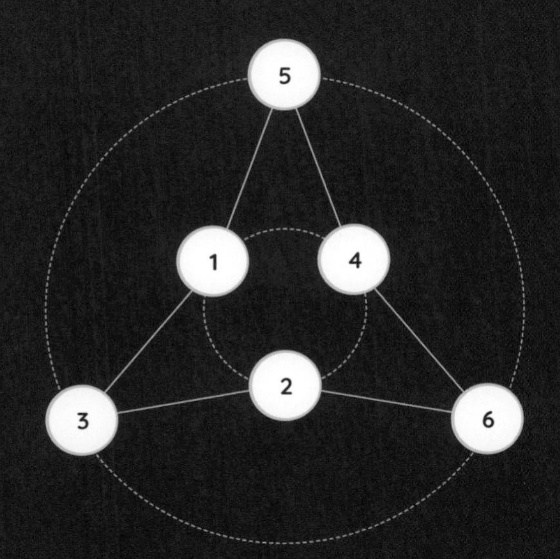

Why? Because the sum of the numbers on the outer disk's bits must be double the sum of numbers on the inner disks. (Confirm that that is true before continuing. The reason is that each number on the inner disks counts twice in creating the outer disks.)

This means the sum of all the numbers must be a multiple of 3.
The sum of the integers 1-10 is 55, which is not a multiple of 3.

Experiment with themes that are a little bit edgy... like this puzzle and a later puzzle that has frogs being dropped into bubbling cauldrons. Students (especially the half that have too much testosterone) are often pulled into the drama of a story that is a little bit nasty.

BORING CONVERSATIONS

For example, the four meeple on the top are lining up for a movie. If they go back for the sequel, can all of them line up without two being next to each other twice?

The arrangement for the sequel doesn't work because although red and **teal** are happy, the yellow and **blue** meeples have the same boring conversation with each other as they did for the movie.

SPOILER ALERT!

Below is a solution. The meeples can happily line up for the movie and the sequel without having the same conversation twice.

Unlike human populations, cloning is both legal and commonplace. Meeples of the same color are happy to have conversations among themselves... but only once. Meeples of different colors are happy to have conversations among themselves... but only once.

Eight meeples lined up for a movie and a week later lined up for its sequel. Find a way for them to line up so that no meeple is talking to the same color of meeple on both lineups. Here, the **blues** and **teal** are happy, but the **red** and yellow meeples are talking to each other both weeks.

Here is one solution. That means that (3,2,**2**,**1**) is possible. Not all combinations are possible of course. For example, (4,2,**1**,**1**) is impossible.

Can you find a way for (3,2,**2**,**2**) meeples to meet twice in a circle? The two below do not work because:

- the red clones are talking to each other in both circles.
- the yellow and blue clones are talking to each other in both circles.

SPOILER ALERT!

Above are (3,2,**2**,**2**) meeples meeting twice in a circle.

- What groups are possible for an 8-meeple circle?
- A 10-meeple circle?
- Other circles?

A square classroom of sixteen meeple students (4,**4**,4,2,2) want to meet twice. Each meeple talks to the meeples left, right, up and down, but not diagonal. Can you find a way for them to meet twice? The arrangement of students below is not a good way to begin. 🙂

What are the fewest number of meeples that can meet three times in a circle? **You'll definitely need more than five colors.**

SPOILER ALERT!

The classroom of meeple students (4,**4**,4,2,2) can meet twice in this way. There might be other ways, but I think they all look similar to this one.

Seven meeples can get together in a circle three times. We can't recruit fewer than seven because the each meeple must talk to two different meeples each time.

For different sizes of circles, what combinations allow meeples to meet three times? Four times? Etc. What about other shapes?

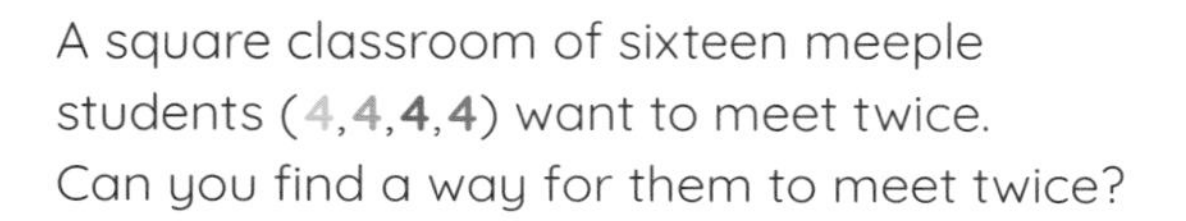

A square classroom of sixteen meeple students (**4**,**4**,**4**,**4**) want to meet twice. Can you find a way for them to meet twice?

This seems impossible. **On this page are some of my beautiful failures.**

They really are beautiful—aren't they?

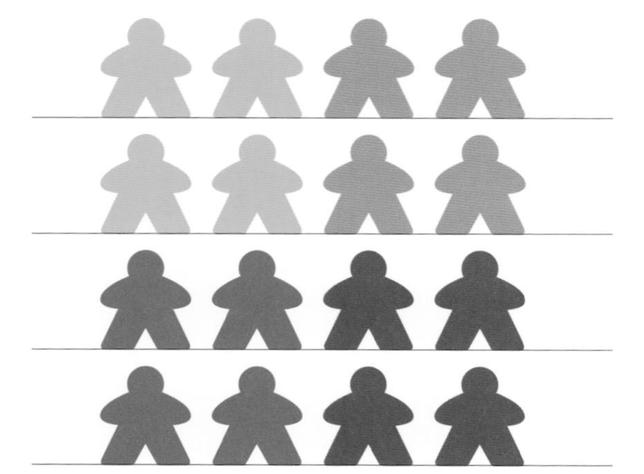

When attempting puzzles, children and adults should usually start with a pattern. **Why?**

1. Because the human brain is good at playing with patterns. Insights are easier to have with a pattern than a random solution attempt.
2. There are usually a lot more random than patterned attempts at a solution. Patterns solve a bigger share of puzzles than they deserve. That means even if your brain were not good with patterns, it is still a good place to start hunting for a solution.

GRAPHENE TRAMPOLINE

Once upon a time in mid-semester, chemist **Richard Smalley** constructed a magical ball. He didn't know it was magical until he peered through it. Then his feet left the floor, he shrank out of his shoes and was sucked inside.

He was stuck in there for a long time. To entertain himself he jumped from carbon ring to carbon ring—using them as trampolines. Soon he began to create puzzles.

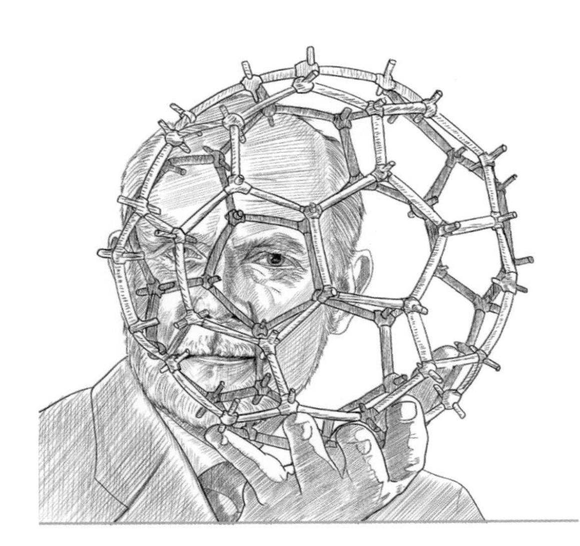

Example: starting at the circle, Richard rolled a 3 and needed to reroll. He then rolled a 4 and jumped in that direction. Some of the other jumps required many rolls, but the outcome was certain. He ended up winning by visiting all trampolines.

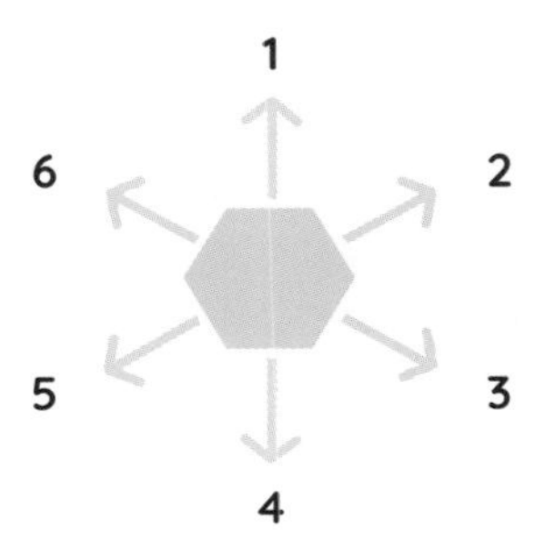

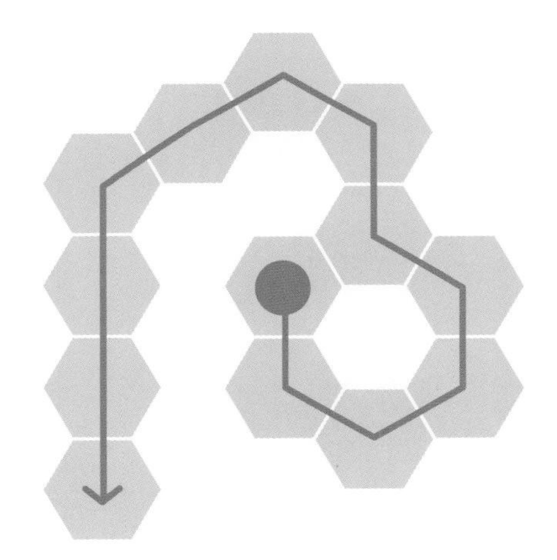

However, starting at the circle, if he had rolled a two, he was doomed to failure.

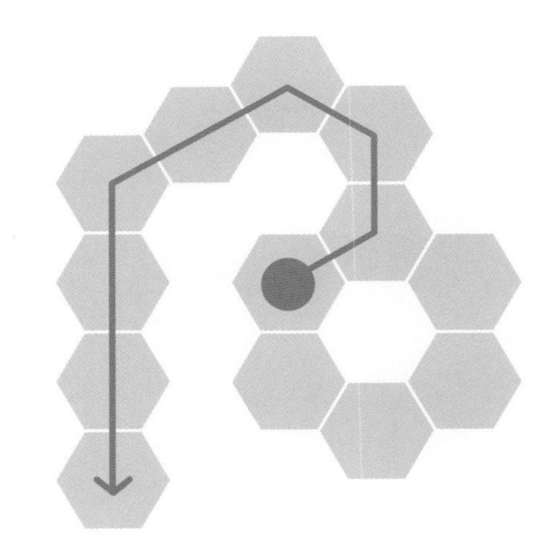

Starting hexagons where success and failure are both possible are colored yellow.

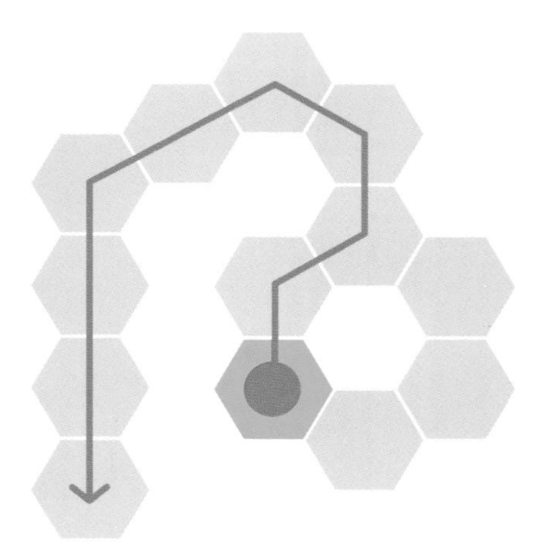

Starting hexagons that always fail are colored orange. It does not matter how Richard rolls. He will always fail starting here.

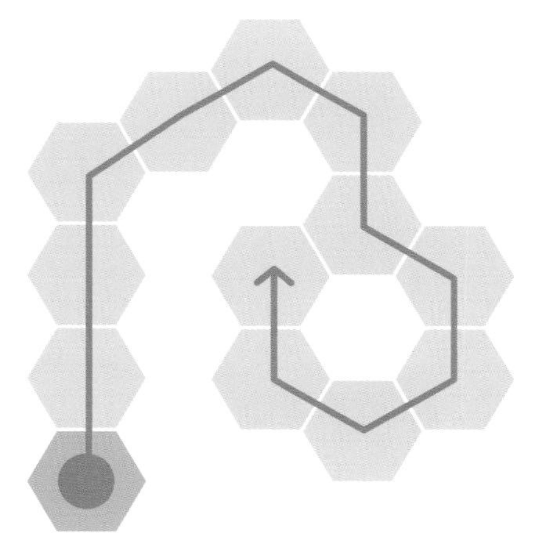

Starting hexagons that always succeed are colored green. It does not matter how Richard rolls. He will always succeed starting here.

Google "Mathigon" for tools to help you make your own puzzle.

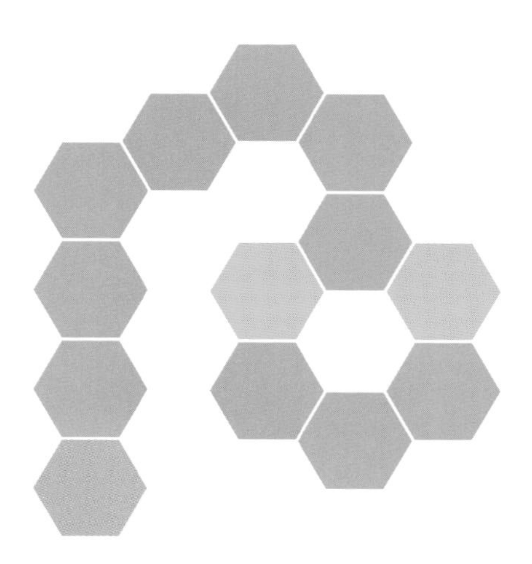

On the following pages color each hexagon yellow, orange or green.

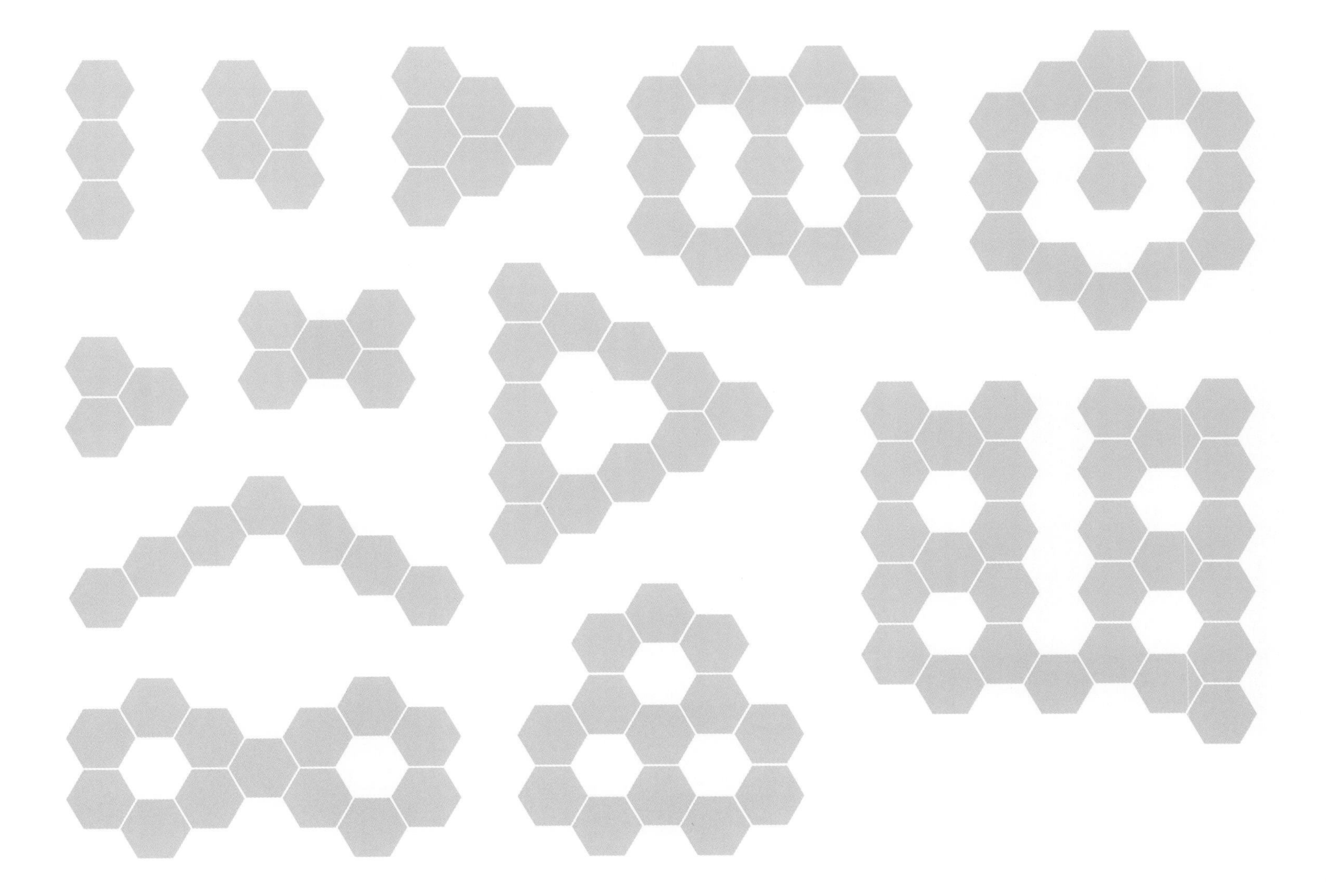

ABCDEFGHIJ
KLMNOPQR
STUVWXYZ

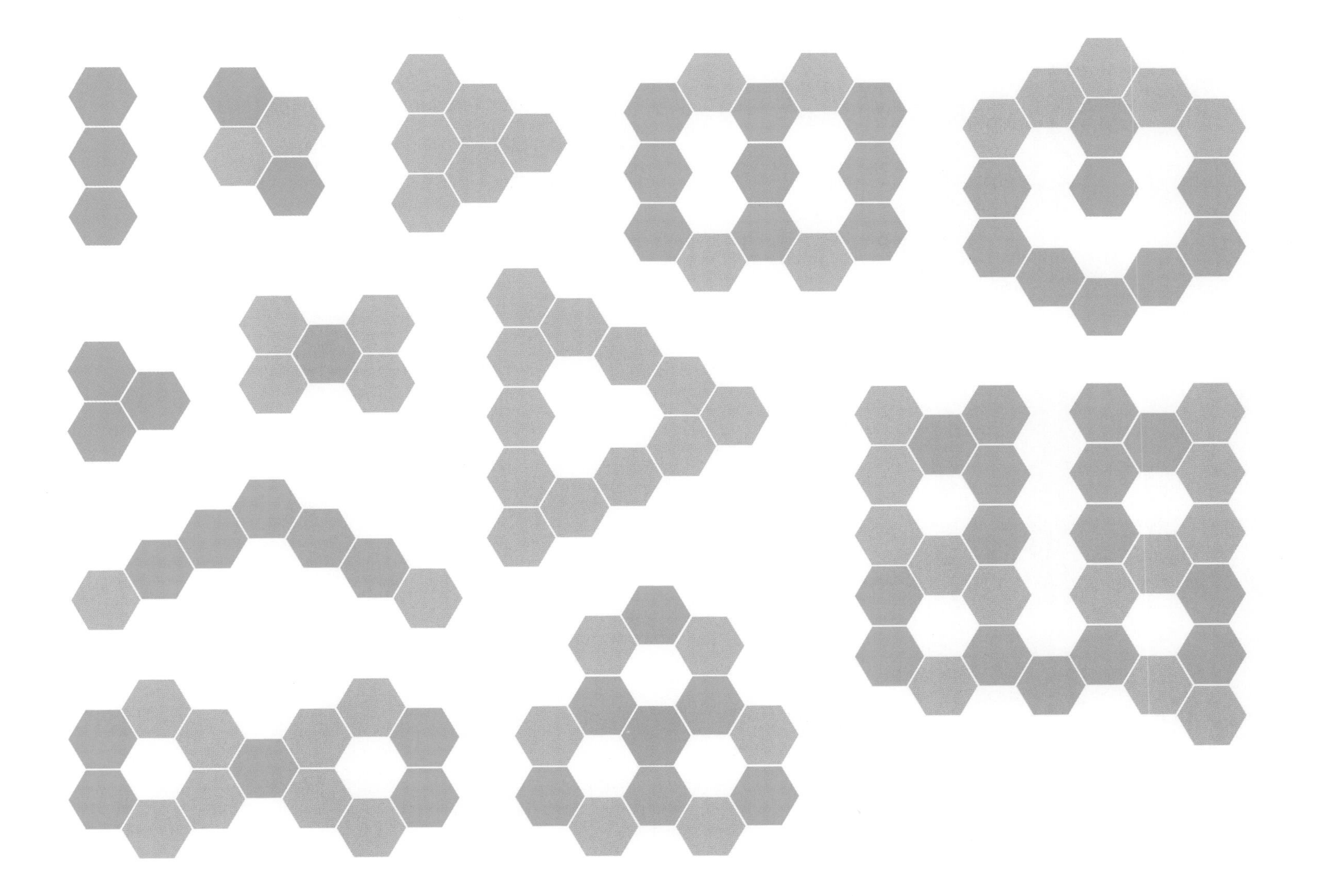

The twenty six letters of the alphabet on the last page do not look fun for most mathematicians, but you'll find that many children who find the abstract puzzles intimidating are drawn to these familiar shapes.

Up till now, this has not been an infinite puzzle. Now we are going to make it one, but only for you pickle people! Find the probability of winning from each starting hexagon. The highest probability of success is 1. What is the second highest probability of success from any hexagon in any puzzle? Find the third through fifth highest probability of success. Keep going...

SPOILER ALERT!

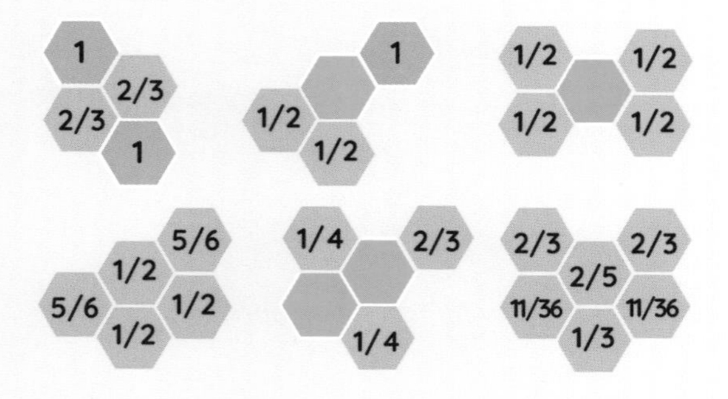

The highest probabilities of success that I have found are 1, 5/6, 29/36 and 3/4. After that I'm stuck and I'm not even sure these are the best.

Now substitute these hexagonal puzzles with any planar graph where the probability of moving from node to node is uniformly chosen from among the connected, unvisited nodes. **The highest probability of success is 1.** What are the next highest probabilities of success?

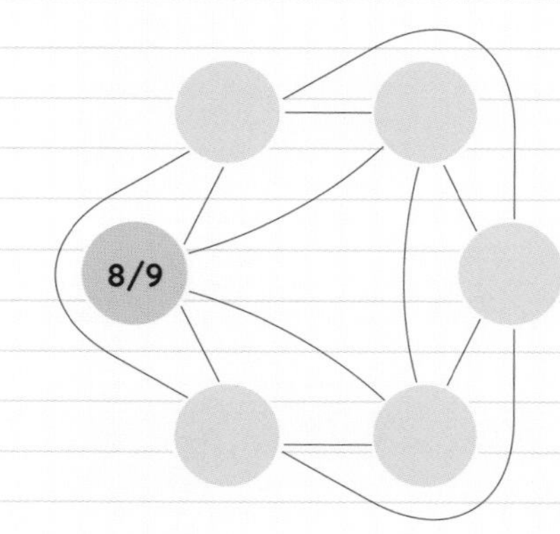

The second highest I found was 17/18. The third highest was 8/9 from the octahedral graph above. **I'm not confident that these are the best.**

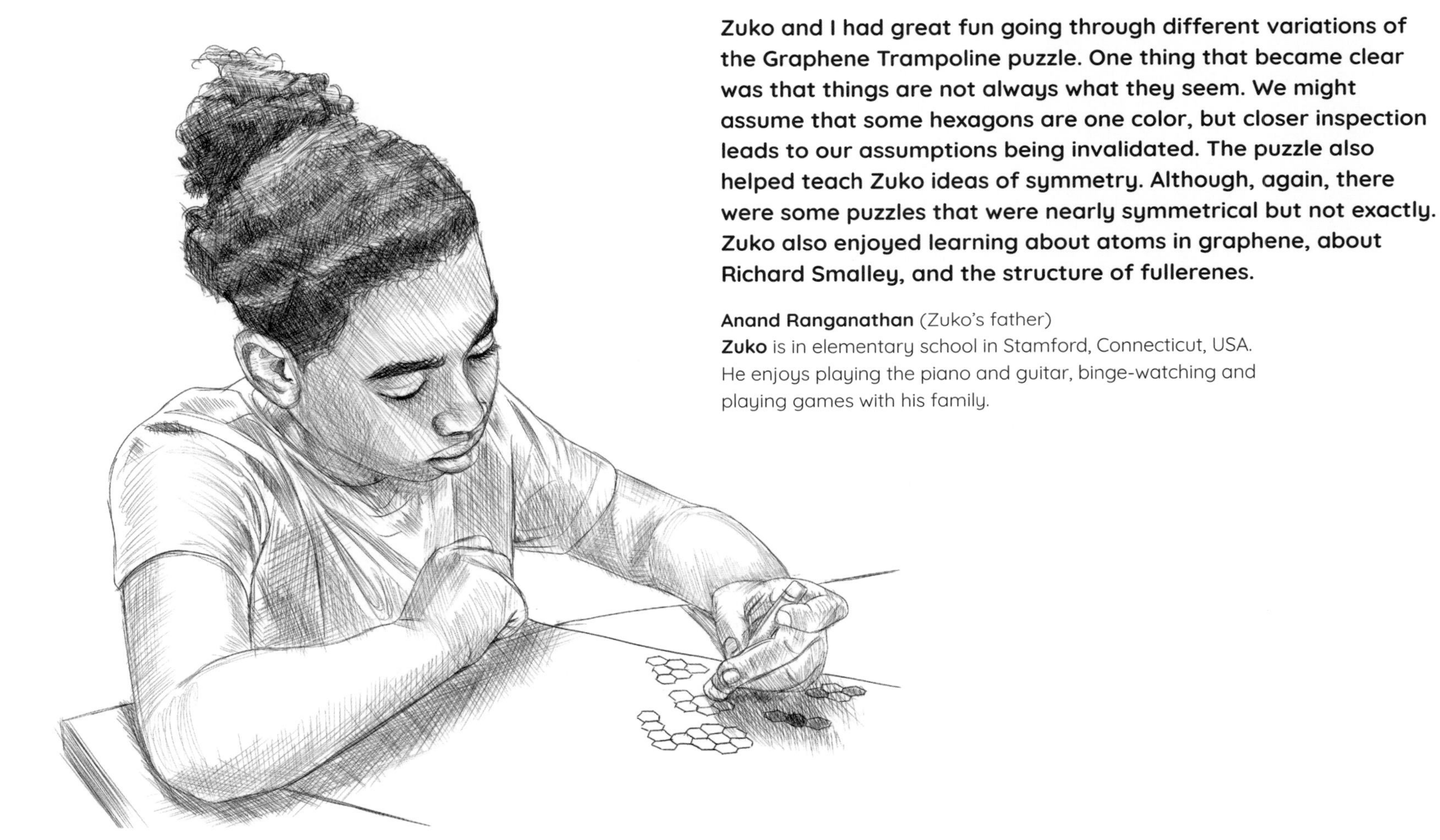

Zuko and I had great fun going through different variations of the Graphene Trampoline puzzle. One thing that became clear was that things are not always what they seem. We might assume that some hexagons are one color, but closer inspection leads to our assumptions being invalidated. The puzzle also helped teach Zuko ideas of symmetry. Although, again, there were some puzzles that were nearly symmetrical but not exactly. Zuko also enjoyed learning about atoms in graphene, about Richard Smalley, and the structure of fullerenes.

Anand Ranganathan (Zuko's father)
Zuko is in elementary school in Stamford, Connecticut, USA. He enjoys playing the piano and guitar, binge-watching and playing games with his family.

WILLIE WIGGLE WIGGLE WORM:
TOOTHPICK MAZES

Willie Wiggle Wiggle Worm is a good student, but sometimes is squirmy. Use the fewest number of toothpicks so that Willie has exactly one way to curl up. Does this work?

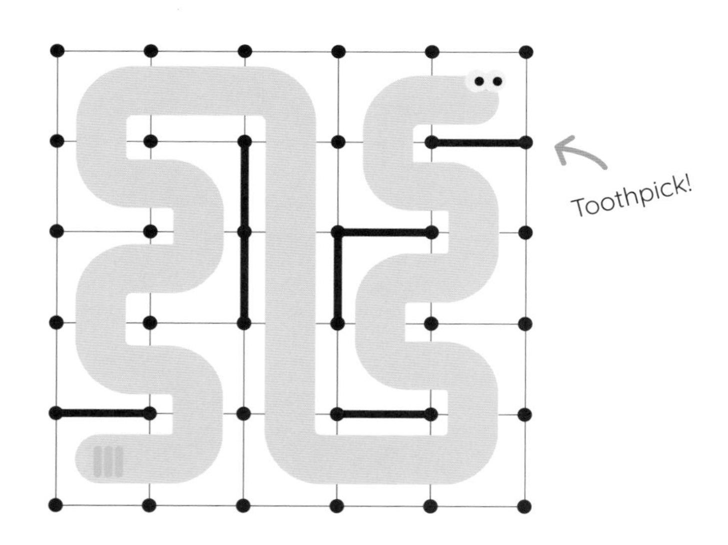

No - because if Willie finds one solution like the one above he can just switch his head and tail to find a second way to curl up! Solutions always come in pairs.

So let's say that your objective is to find the fewest number of toothpicks that has exactly one pair of solutions. This set of seven toothpicks is not enough because there is another pair of solutions.

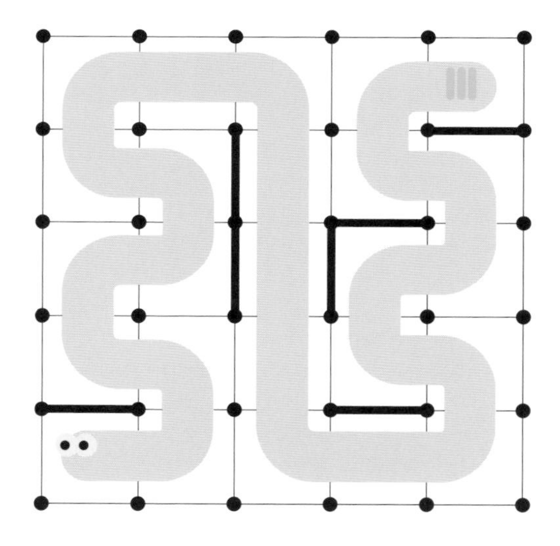

Find the other pair of solutions.

SPOILER ALERT!

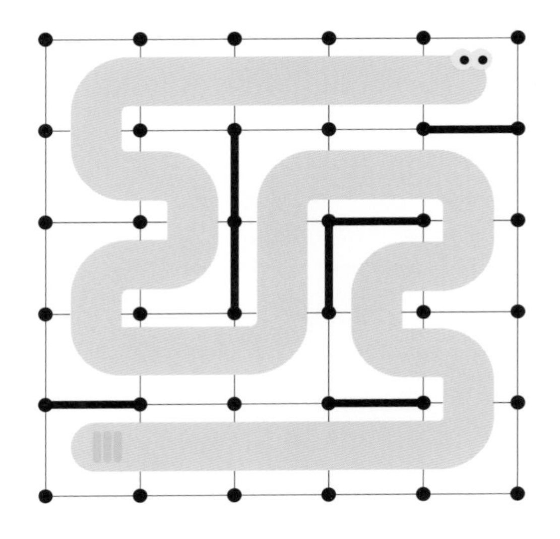

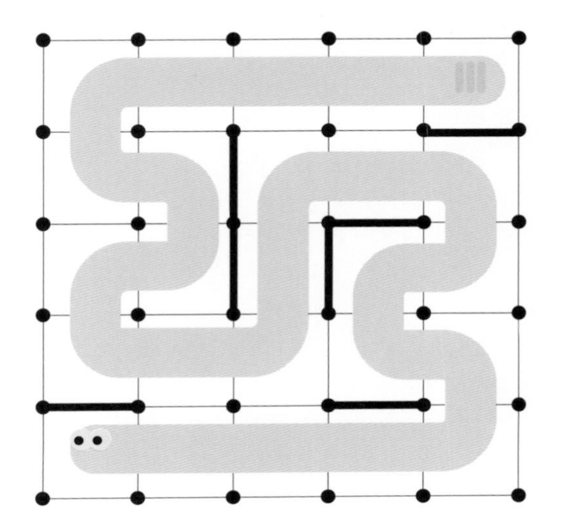

Here is the other pair. Try to find the minimum number of toothpicks needed to ensure just one pair of solutions exist in 5x5 and 6x6 cartons. **I'll share the best solutions I've found.** For the 6x6, I failed to find a solution using only six toothpicks.

I'm guessing seven is the fewest possible for 6x6. I won't tell you how well I did for 5x5 until you try. Smaller cartons are also interesting. 3x3 cartons only need two toothpicks. Do any of these 4x4 toothpick arrangements yield only a single pair of solutions?

SPOILER ALERT!

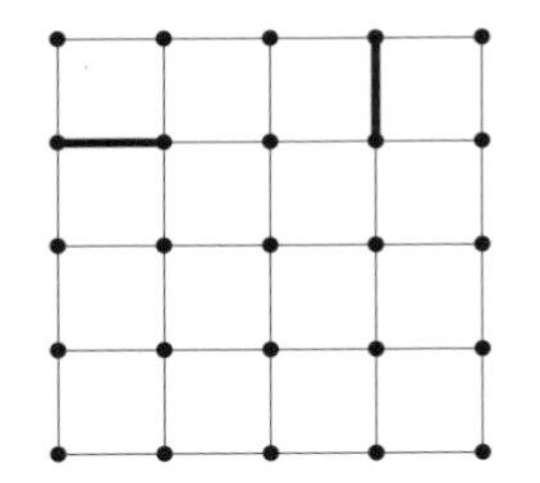

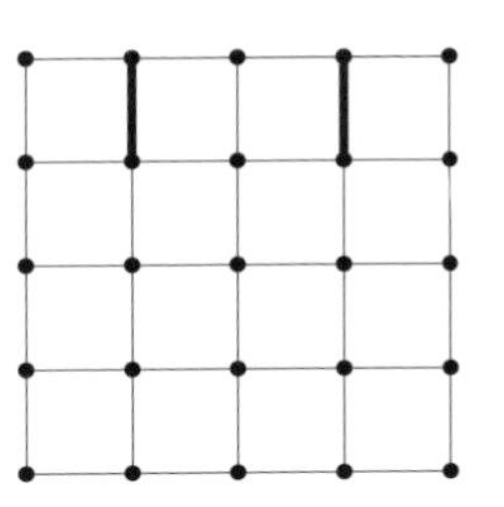

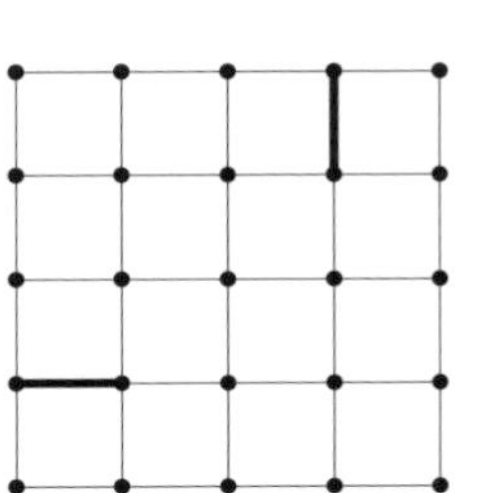

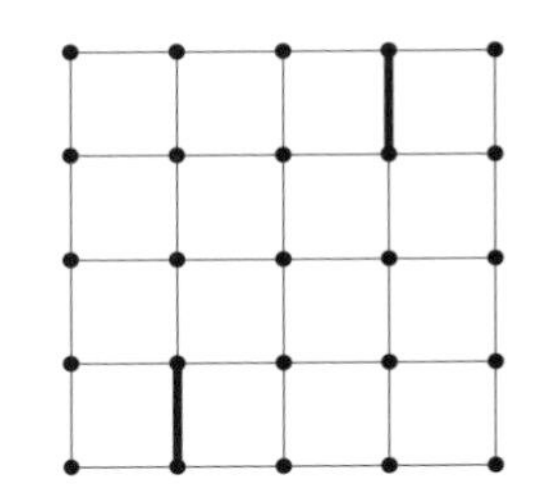

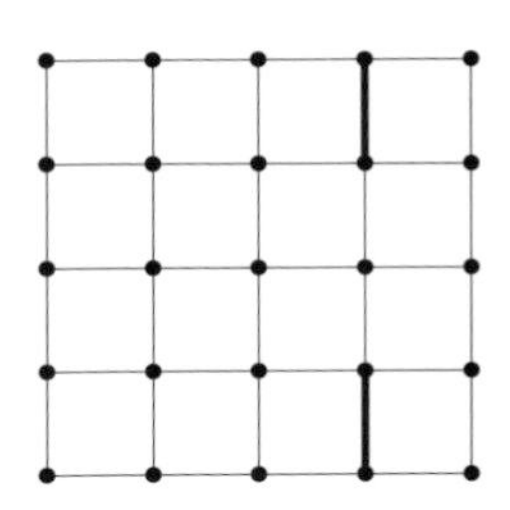

The fewest number of toothpicks I needed in a 5x5 carton was four. The six toothpick idea in the 6x6 carton fails, but it is tantalizingly close! There are two pairs of solutions. 🙁

From now on, I'm not going to show both solutions of a pair—it is not really exciting to see which end is the head and which is the tail.

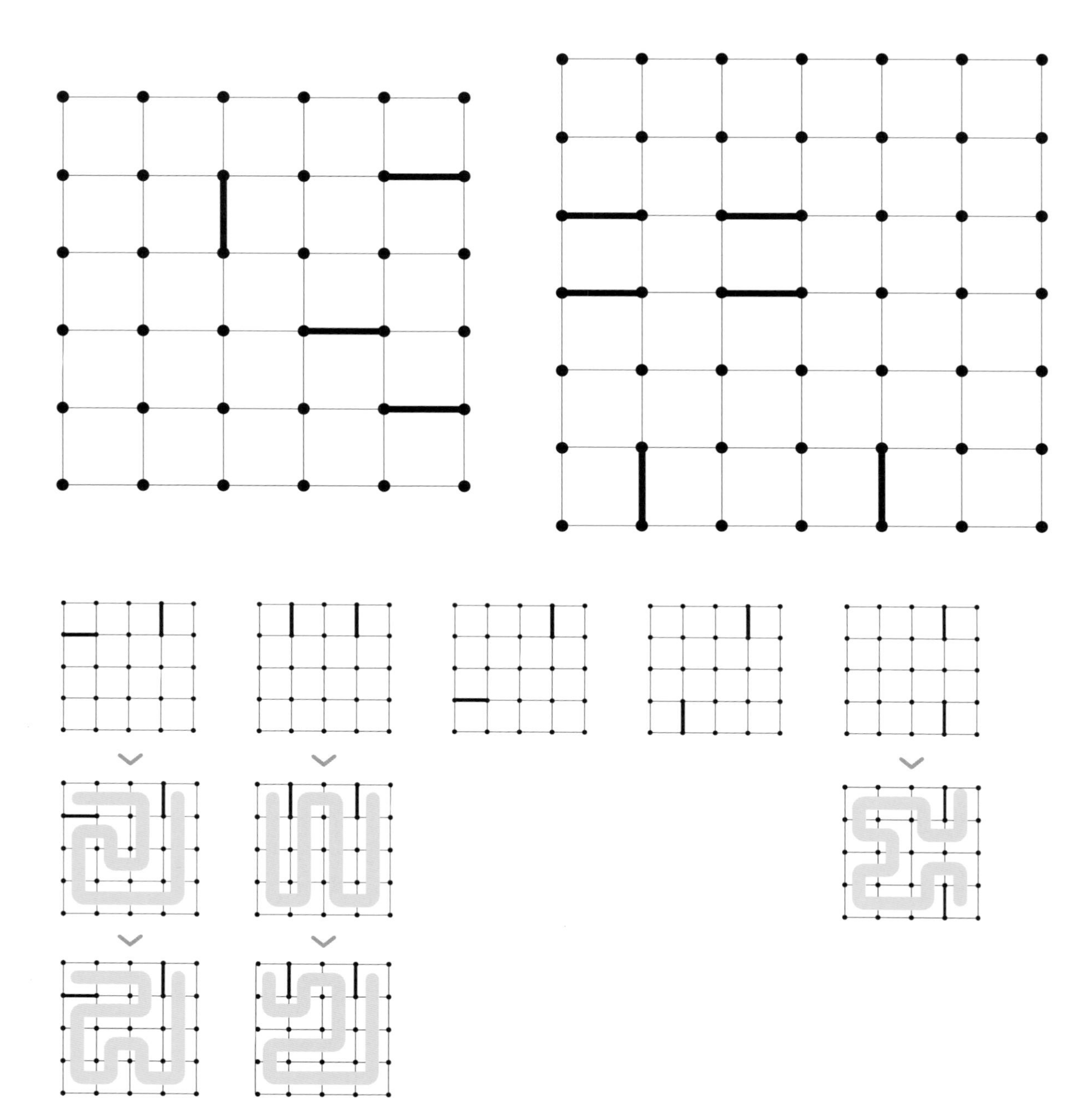

On the right, we see the answer to the 4x4 question. The first two have got more than one way for Willie to curl up. The next two are impossible. **The last one works!** Two toothpicks are sufficient to ensure that Willie only has one pair of ways to curl up.

There are other ways to place four toothpicks in a 5x5 that result in an answer. So if you got a different solution, celebrate. 🙂 If you got a solution using fewer toothpicks—wow! I would be really impressed, but make sure you check your work!

The Infinite Pickle would not be liberatingly honest if it did not include an epic fail. Here it is. For the months of writing the book I thought that Willie had only one pair of ways to curl up in the 6x6 on the right. I was wrong. If you find a six-toothpick solution with only one pair of ways to curl up—wow! I would be both skeptical and impressed!

The fewest number of toothpicks I've found for a 5x5, 6x6, 7x7 and 8x8 are 4, 7, 10 and 11 respectively.

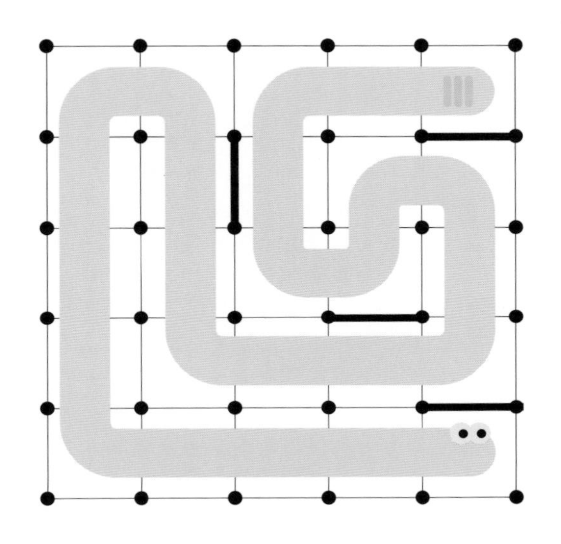

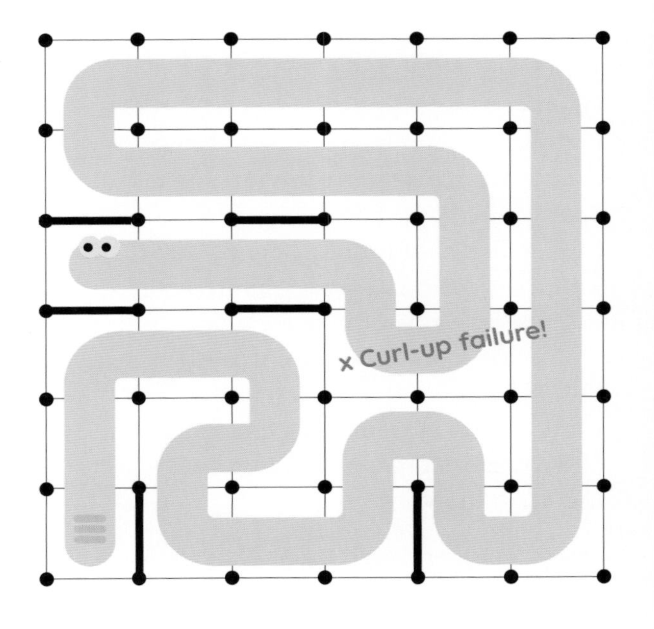

Students can propose designs and other students can try to break them by finding multiple ways for Willie to lie down. It is fascinating to see children naturally specialize into maze-designers and maze-breakers.

Willie can't suck his thumb to fall asleep so sometimes he curls up in a loop and sucks his tail.

- What are the fewest number of toothpicks needed so that he has only one way to do this in a 2x2, 3x3, 4x4, 5x5 or 6x6 carton?
- What about rectangles of width 2?
- Width 3?
- Width 4?

SPOILER ALERT!

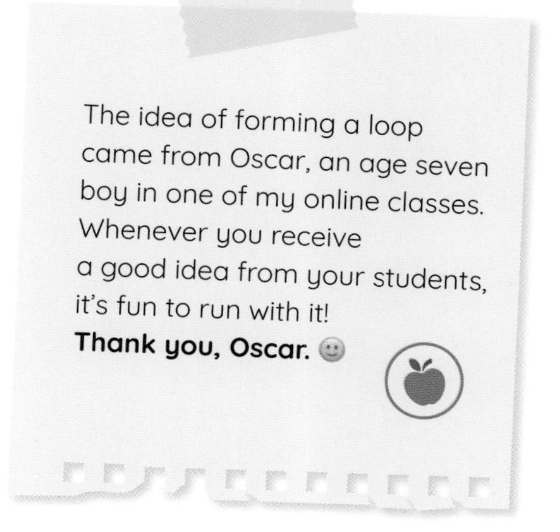

This is one of my favorite puzzles. I didn't think of it as an animal. I just thought about it being a loop. Doing a loop is a different puzzle than just doing a line. When it's a loop it's often more challenging. I'm not sure why it's more challenging, but it just is. Once you make the dimensions all four or more, or really, starting at a 3x4, it gets fun. It's kind of a combination of geometry and logic and maybe there's a little graph theory in there.

Oscar was born in New York City and now lives in California. At age seven he is already a mathematician in all respects except having a degree. He is self-taught.

First, let's acknowledge how unbelievably boring it is for Willie to lie down in a loop for any rectangular carton that has width 2. No toothpicks are required.

A 3x3 square and all rectangles with an odd number of squares are a bit more difficult. In fact, they're impossible. **Why?**

Consider the coloring of Willie if he lies down in a carton freshly painted in a checkerboard pattern. When the paint dries, Willie will end up being striped in a glorious alternating pattern—a pattern with an even number of stripes! Willie is pretty sure there's a contradiction in there somewhere.

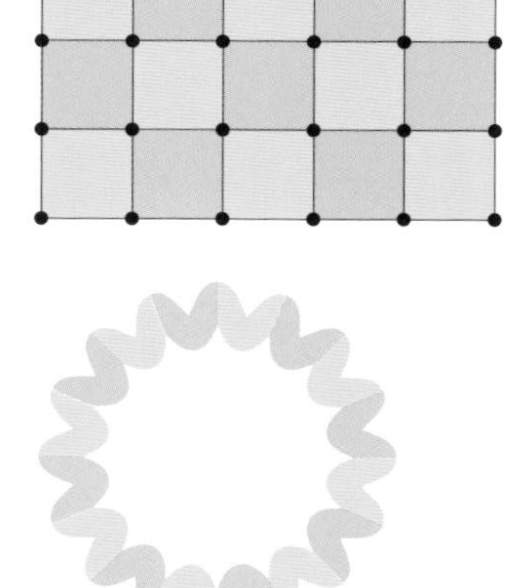

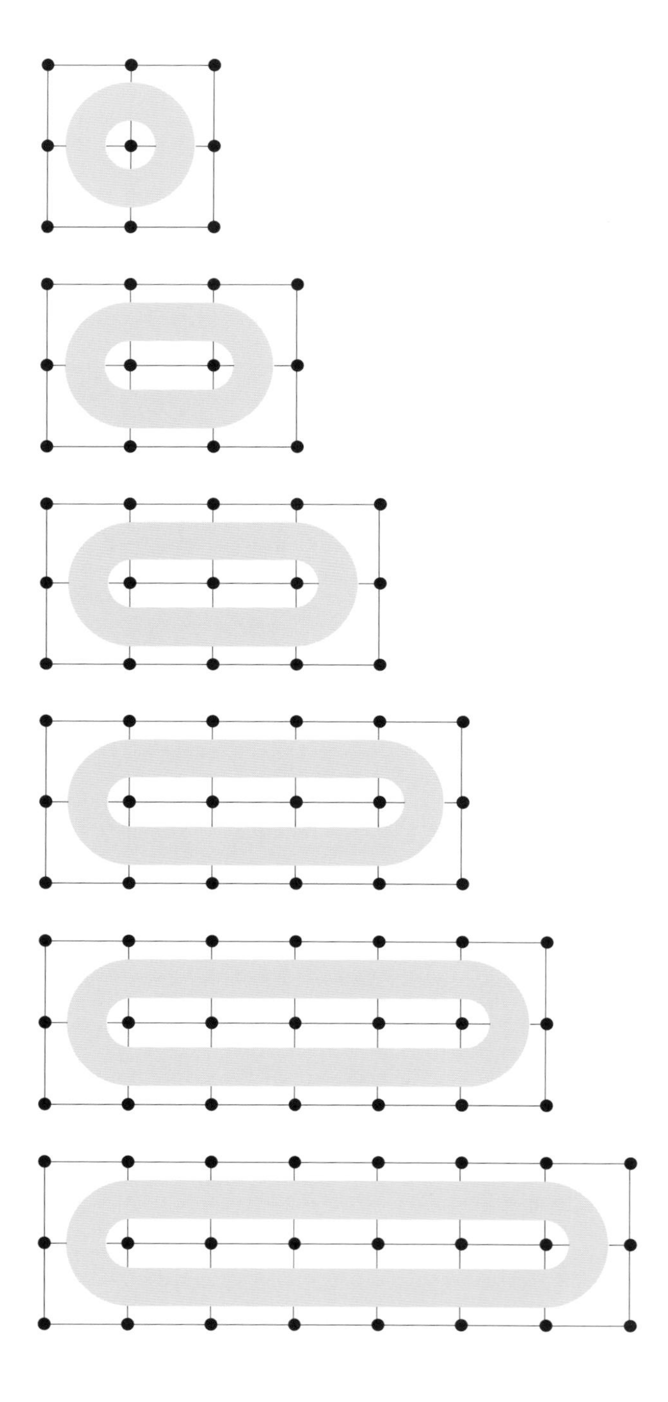

A 3x4 rectangle needs one toothpick; a 3x6 requires two toothpicks; a 3x8 (below) requires three. The toothpicks can be all on the left or all on the right or a mix. This pattern continues with 3x10, 3x12 etc.

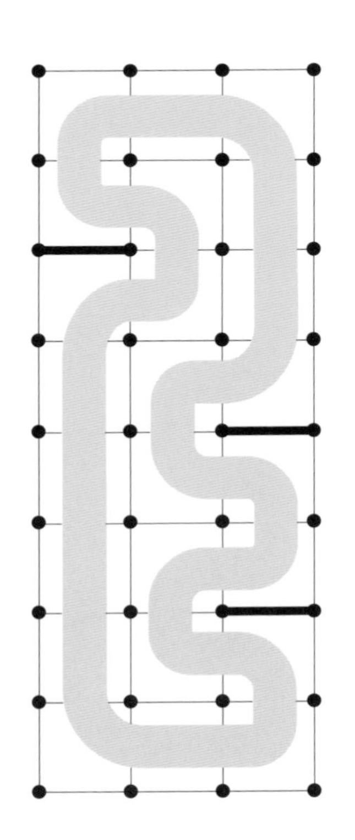

A 4x4 square requires two toothpicks. The 6x6 requires at most six toothpicks. Is this the best possible? I don't know.

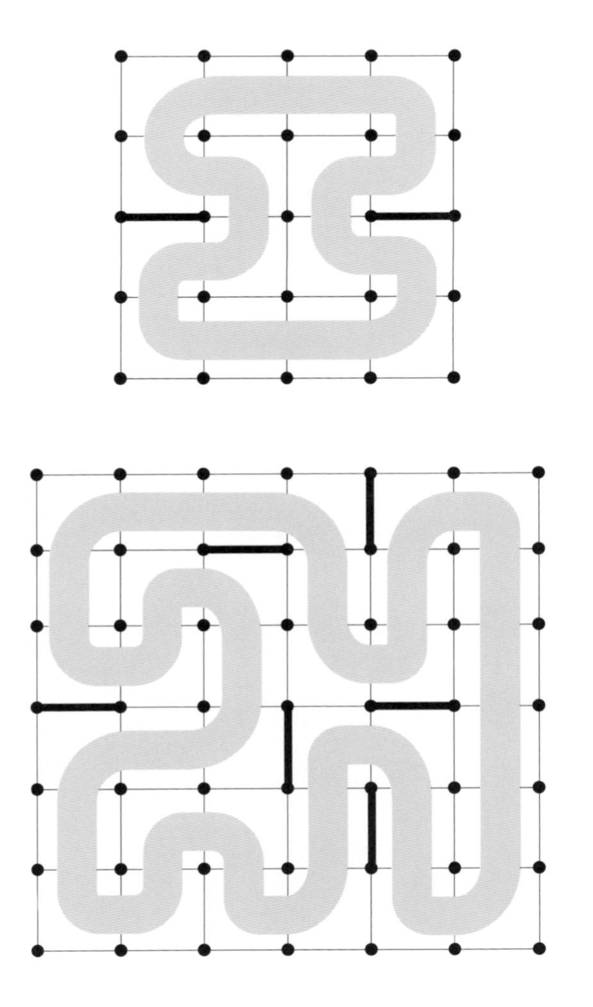

The 4x10 rectangle requires only four toothpicks! **This pattern can be extended for longer rectangles of width four.**

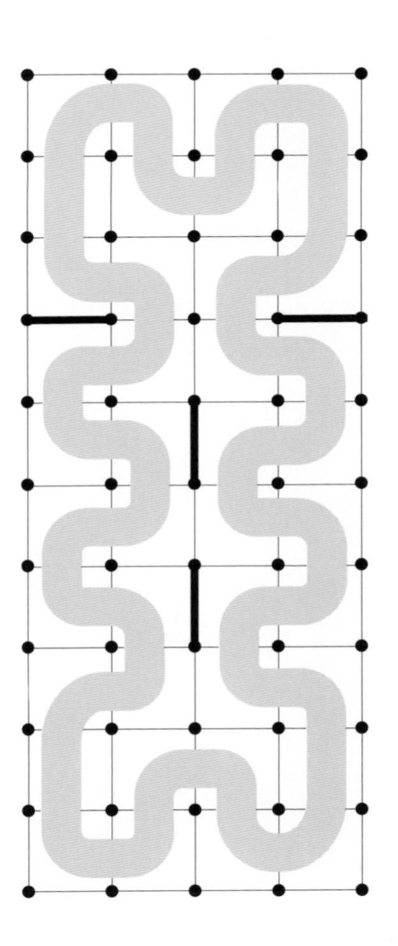

PINOCCHIO'S PLAYMATES

When Pinocchio gets to school, he has an unfortunate ability to turn some of his truth-telling classmates into liars.

His first class has 16 students with desks in a 4x4 classroom.

Pinocchio's teacher asks the students to raise their hands only if they are seated beside (left-right-front-back) exactly two like-minded students. **What do the students do?**

- A truth-teller raises their hand only if they are seated beside exactly two truth-tellers.
- A liar lies. If they are seated beside exactly two other liars they do not raise their hand. Otherwise they do.

When the teacher asks the question, all children raise their hands! Out of the 16 students, how many are liars and how might they be seated in the 4x4 classroom?

SPOILER ALERT!

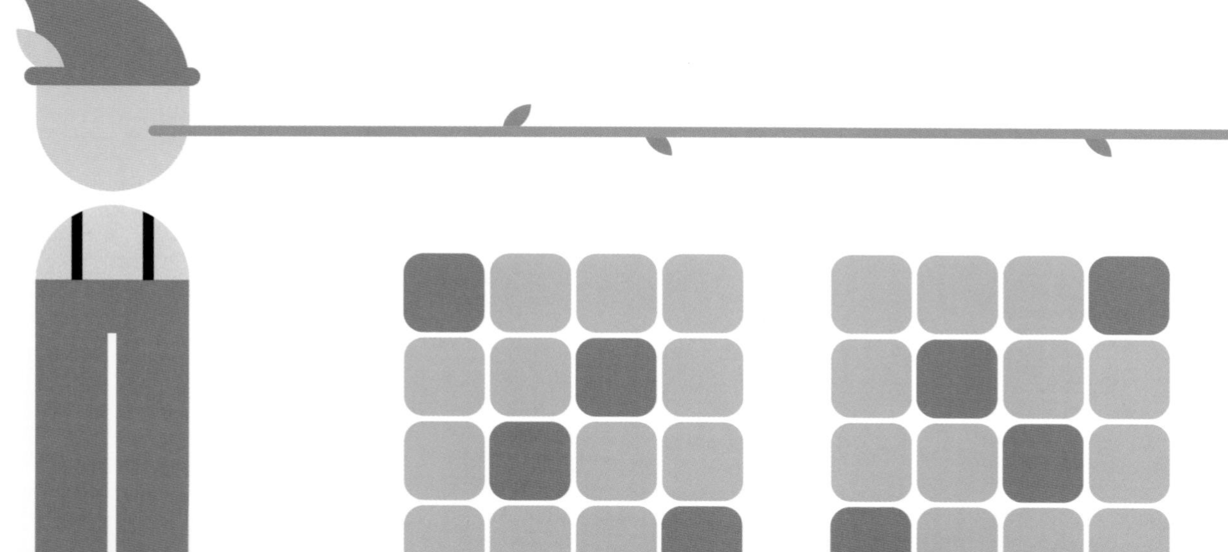

There are four liars.
The class is in either one of the patterns on the left.

The twelve truth-tellers are green. The liars are **red.**

Pinocchio's next classroom is 5x5.
The teacher asks the same question and gets the same result! All students raise their hands. How many students are liars and how might they be seated?

This doesn't work. Why? It doesn't work because the student in the center is a liar. He is seated beside two other liars so if he raised his hand he would be telling the truth.

There is only one solution for the 5x5 classroom. There are exactly 9 liars and 16 truth-tellers.

Getting ready for recess, the students sometimes line up in the corridor. It is just wide enough for two students to stand beside one another. For any even number of students, find all possible rectangular configurations of liars and truth-tellers so that all students would raise their hands when asked if they are beside exactly two like-minded students. This corridor exploration is by Taiwanese educator **CinJhih Zeng.**

SPOILER ALERT!

When students line up in the corridor, there is one way to do it—except when the number of students is a multiple of three. Then there are two ways.

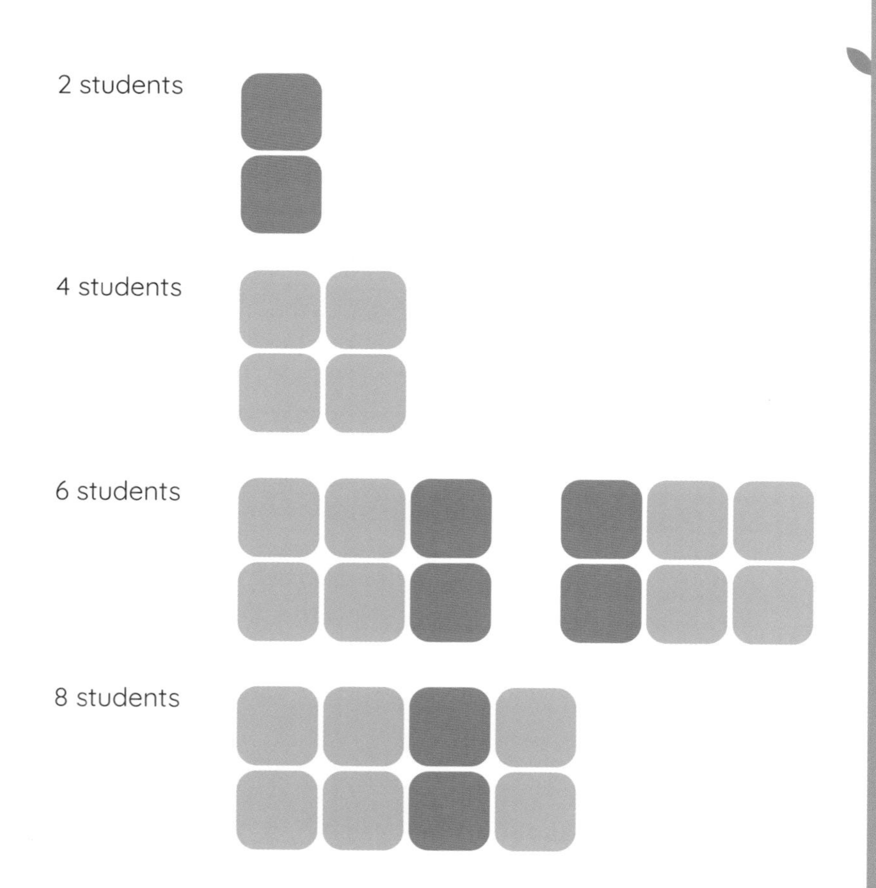

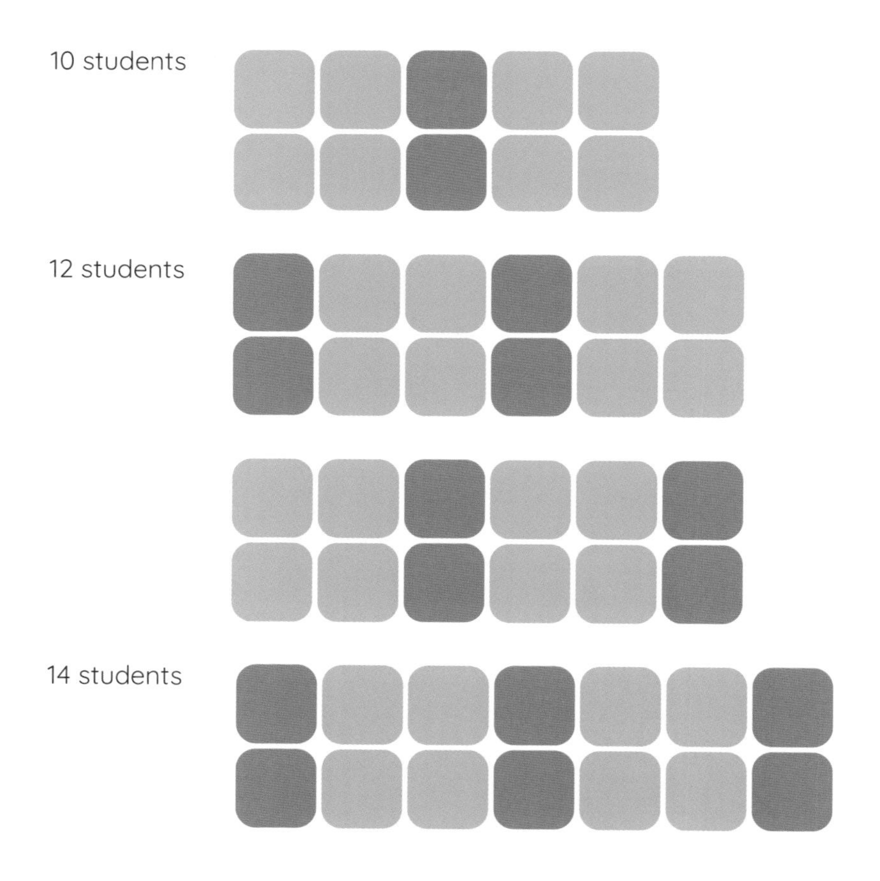

Which classrooms can have all students raise their hands? Is it possible with 5x6, 5x7, 6x6 or 6x7 rectangles? The 6x7 rectangle below does not work because some students in the second row do not raise their hands.

6x6, 7x7, 8x8, and 9x9 classrooms all raise their hands. In each case, do you know the number of truth-tellers in the class?

SPOILER ALERT!

Earlier this week I used Pinocchio's Playmates after my class played some Truth Teller and Liar Problems. It was great.

Instead of only looking at squares and chunky rectangles as in Gord's original puzzle, I encouraged my kids to find a pattern with long skinny rectangles of width 2. I told Gord about this and he's included it—he's also extended the puzzle to include frieze patterns.

Cin Jhih Zeng teaches math with origami and puzzles. He creates mathematics curricula, designs his own puzzles and helps run the Math Board Game Camp in Taipei.

Here are solutions for the 5x6, 6x6 and 6x7 rectangles. **I think the 5x7 rectangle is impossible.**

5x6

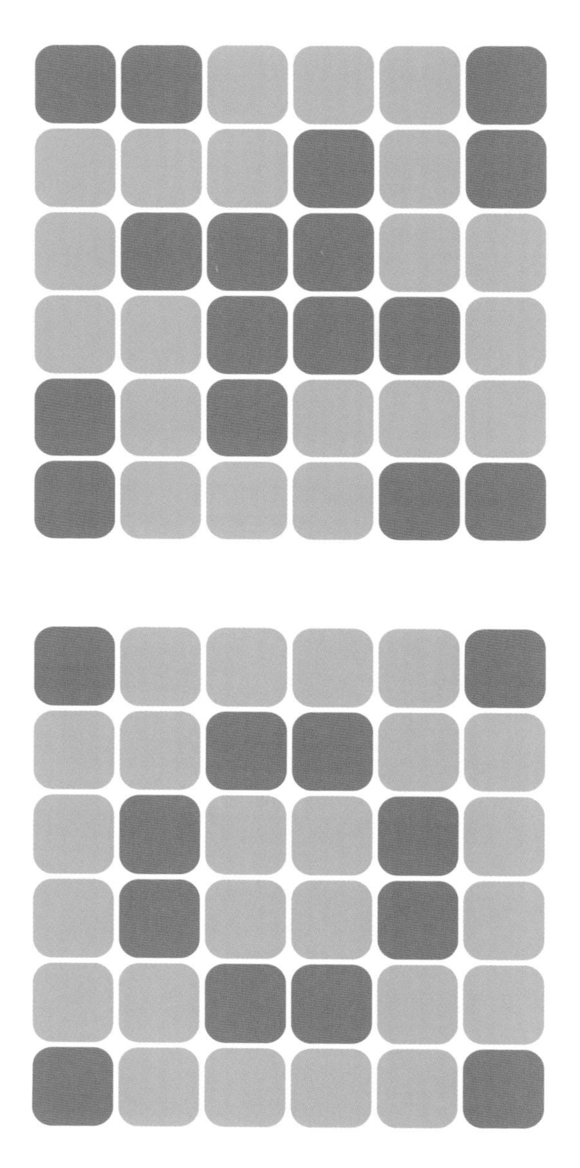

6x6

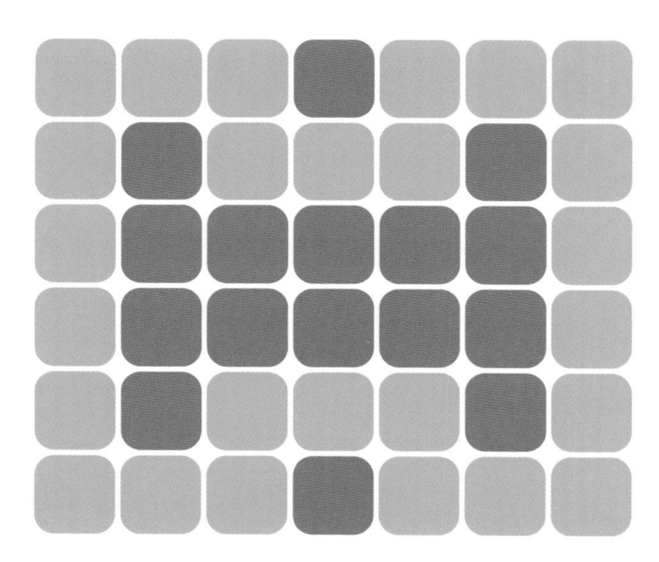

6x7

Feed your class impossible problems on a regular basis.

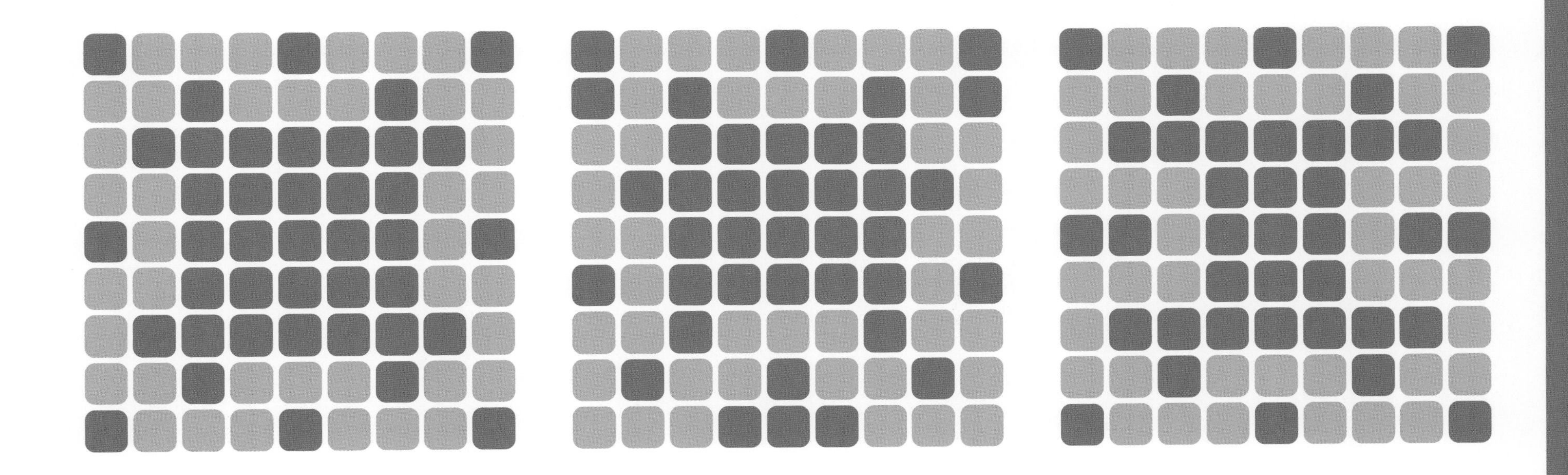

The 6x6 classroom has either 20 or 24 truth-tellers. The 7x7 classroom has 28 truth-tellers. The 8x8 has 32 truth-tellers. All larger squares can have different numbers of truth-tellers. The 9x9 classroom above has between 40 and 44.

When Pinocchio graduates, he finds that his skills in truth manipulation are highly valued. According to Pinocchio the Politician, an infinite number of people lined up to hear him speak. If the corridor is two wide, find some periodic patterns that make Pinocchio's claim possible. **What periods are possible?**
SPOILER ALERT!

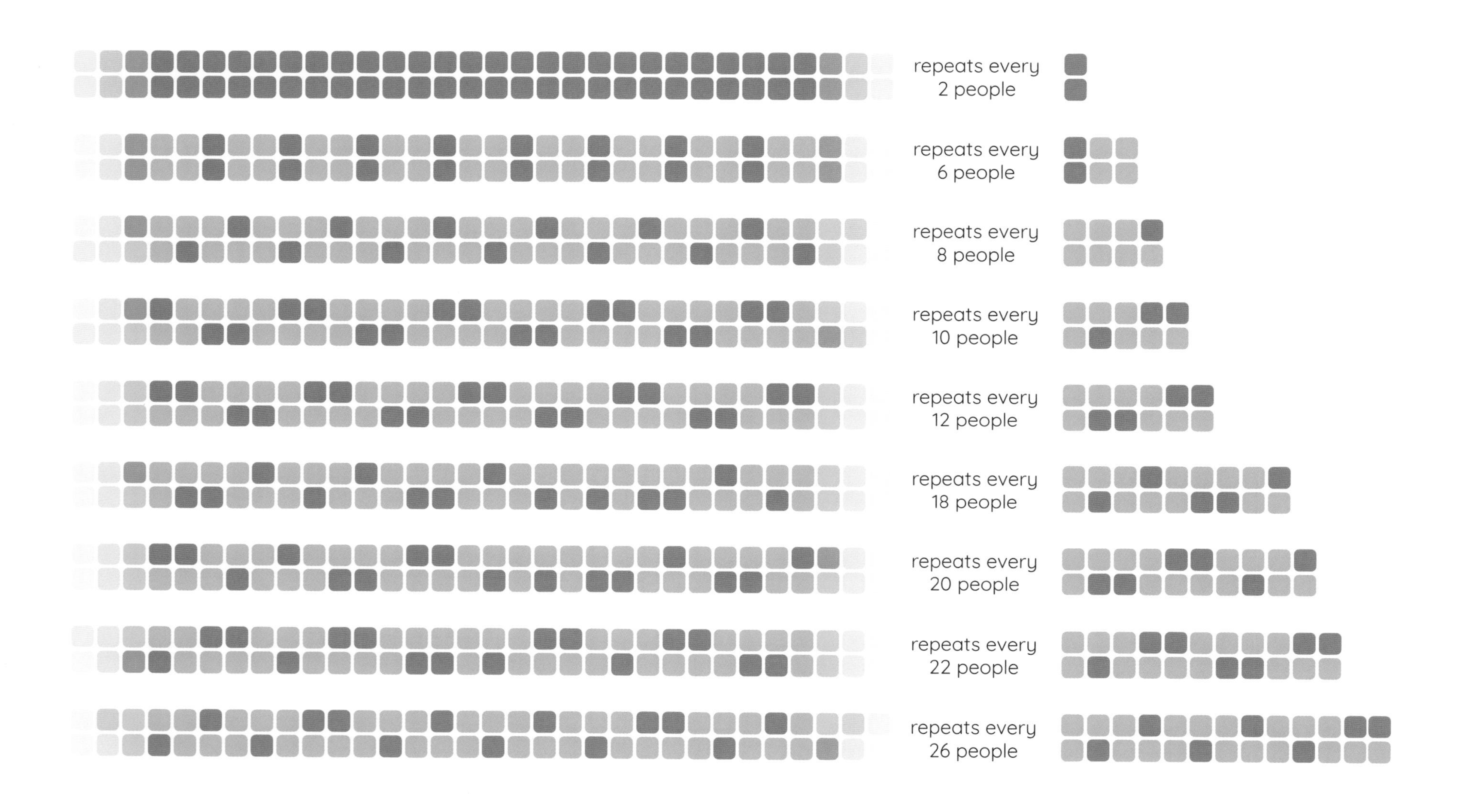

We can organize the solutions according to how frequently they repeat. The bottom solution repeats every 26 people. Find how an infinite number of people could line up if the corridor was 3, 4, 5, 6 or 7 wide. One of these widths may be impossible to solve—I have not found a solution. Oh! One more rule: We must reject solutions—like the top one—that have a row of all liars. **Liars know they have to be more clever than that!**

Periods 4, 14, 16 and 24 seem impossible, but every even number after 26 is possible.

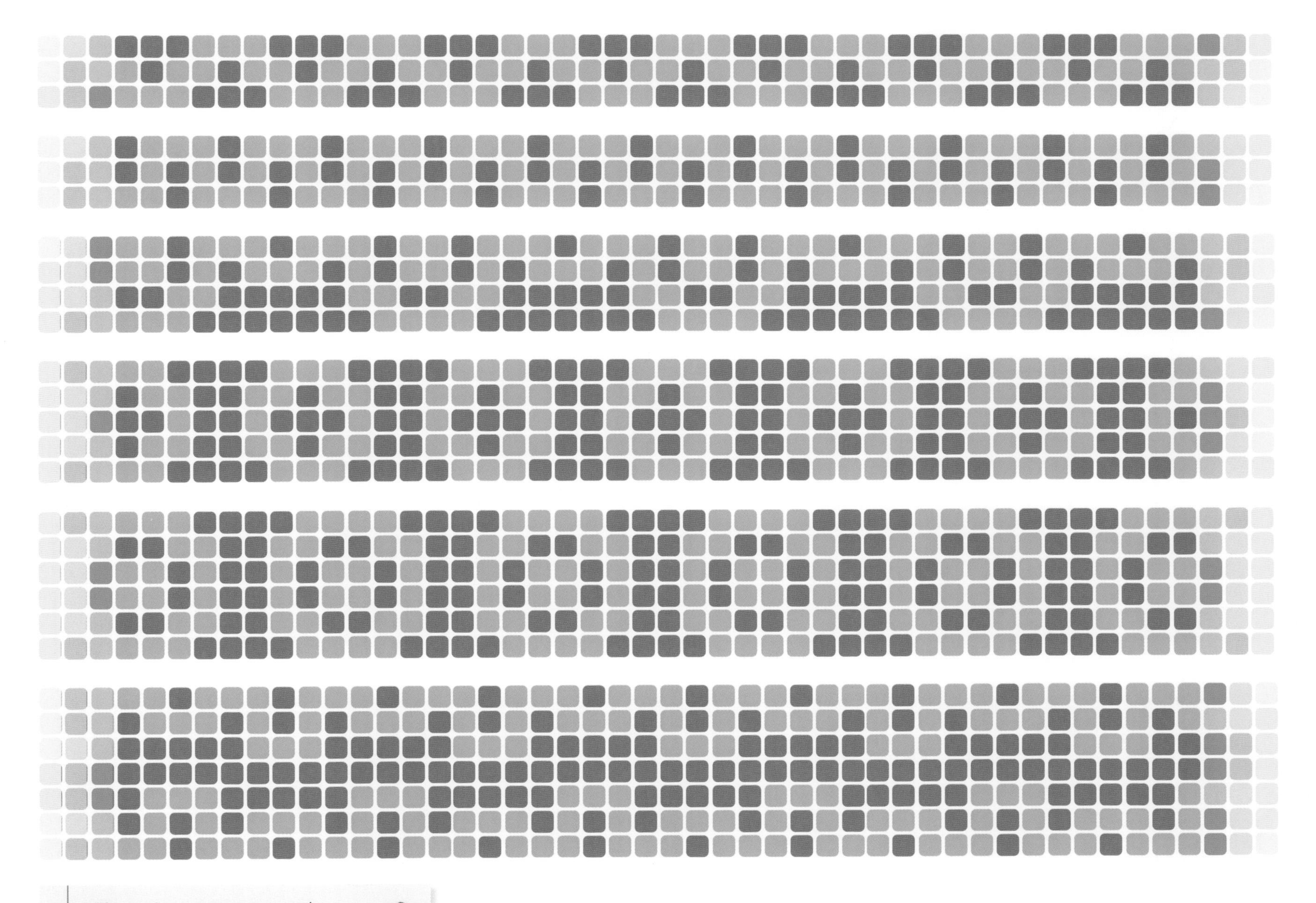

These two pages are only for people geeking out.

I have not found a solution for a corridor of width 7. This one fails because it has a row of liars.

What patterns of truth-tellers and liars would make these little 2x2 classrooms work? As you can see—not all students are raising their hands:

This way to introduce Pinocchio's Playmates is by **Vinay Nair,** co-founder of Raising A Mathematician Foundation in the suburbs of Mumbai, India. Below are the some answers, with green being truth-telling students. The rightmost 2x2 square with no hands raised is impossible.

Here are some of Pinocchio's playmates sitting in a 3x3 classroom. What seating patterns could result in these hands raised?

Here are some seating patterns that could have created those patterns of raised hands:

Did you notice that asymmetric seating patterns sometimes result in a symmetric pattern of raised hands? Is the opposite ever true? Can a symmetric seating pattern ever result in an asymmetric pattern of raised hands? No. This is Pinocchio's upside-down entropy! Things can get more organized, but never less organized.

For a classroom of n students there are 2^n ways for liars and truth-tellers to sit and 2^n patterns for hands to be raised. One consequence of Pinocchio's upside-down entropy is that not all patterns of raised hands will be seen.

What fraction of raised-hand patterns will be impossible? Which might you expect to be the most common?

Imagine that after a 3x3 class is dismissed, a new group of Pinocchio's playmates enters the classroom. Truth-tellers sit in seats that hands were raised in the previous class; liars sit in the remaining seats. This repeats class after class. Here is an example what happens:

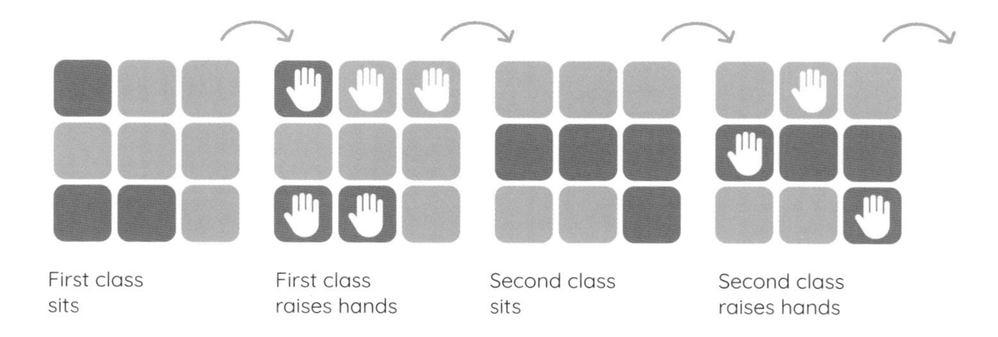

Find all seating arrangements in a 6x6 classroom where the liars and truth-tellers sit in the same chairs class after class.

There are eight 6x6 classrooms that stay the same class after class. These are four. The other four just have the liars and truth-tellers switched.

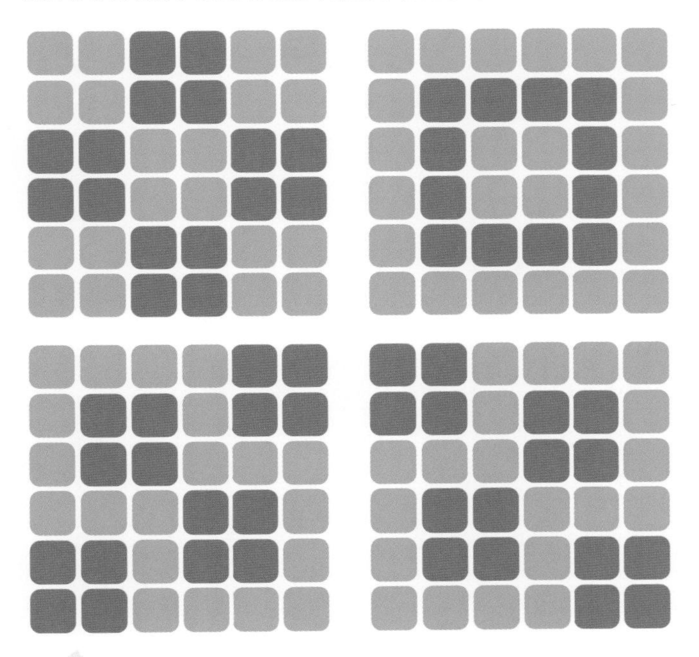

IDEA #1 — Nobody!
Instead of everyone raising their hand, nobody raises their hand.

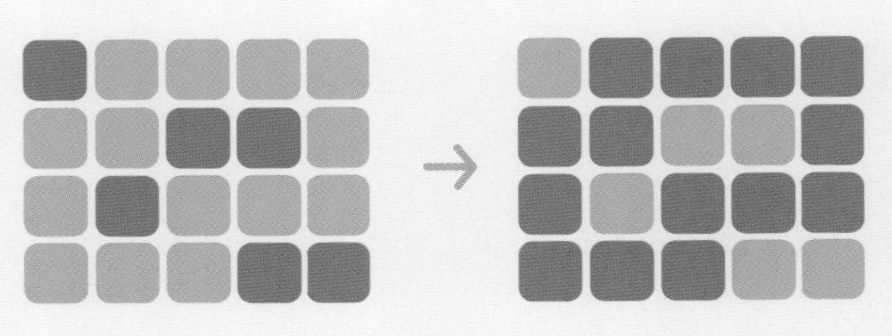

Everybody raises hands → Nobody raises hands

Hypothesis: Switch liars and truth-tellers in a classroom that solves the original puzzle to find a solution to this puzzle.

IDEA #2 — Half & Half!
Instead of all students raising their hands, exactly half raise their hands.

Here are two solutions for a 4x4 classroom:

Hypothesis: All solutions have 50/50 liars/truth-tellers.

Students enjoy being creative—making their own variants to puzzles. Most will be trivial or uninteresting, but getting students to realize that rules are created and can be changed is liberating. This is beneficial even if you decide not to pursue any of the variants.

Here are the best suggestions from ten minutes of brainstorming with two classes of age 10 and 11 students. I've included one student hypothesis in each case. A mark of a vibrant math class is that there are lots of both wrong and right hypotheses, so do not expect all hypotheses to be correct.

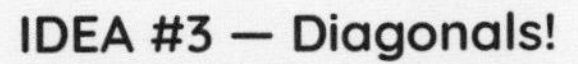

IDEA #3 — Diagonals!
Instead of students raising their hands if they are seated beside (left-right-front-back) exactly two like-minded students, the teacher asks if they are seated diagonally next to exactly two like-minded students. Here is an example of a 5x5 classroom where all students raise their hands:

Hypothesis: All solutions have liars in the corners.

Bubbling Cauldrons

The Macbeth witches are stirring up trouble.

Plop consecutive frogs into a cauldron starting at frog#1, then frog#2, etc. An explosion happens if two frogs in a cauldron add to a new frog plopped in.

It is time to add frog#7 to one of the cauldrons below. Which do you choose?

It is a trick question because an explosion occurs either way: 2+5=7 in the left cauldron and 3+4=7 in the right cauldron.

Start again with frog#1. **How many consecutive frogs can you add to the cauldrons without triggering an explosion?**

SPOILER ALERT!

With a class of young kids, do not start by introducing the rules. Instead, first aim to get emotional engagement. Get as many children as possible to contribute something to the unfolding puzzle.

TEACHER: Anna, into which cauldron do you want to drop frog#1?
ANNA: "The orange one."
TEACHER: "Alonzo,into which cauldron do you want to drop frog#2?"
ALONZO: "The left one again."

The kids still do not understand what is going on. That's okay. Kids are used to learning in this way. Keep going until...

BEATRICE: "Drop frog #7 into the right cauldron."
TEACHER: "Ka-boom! It exploded! You've been gooped, Beatrice! Why? Because two numbers already in the cauldron add to 7...
BEATRICE: "3+4=7"

What child would not like to be gooped? Being "gooped" really means "failure," but by being playful with the word you remove the stigma of failure from the classroom.

TEACHER: "Could you have avoided being gooped by choosing the other cauldron?"
BEATRICE: Yes...no... 2+5 = 7"
TEACHER: "That's right, no matter what you did you would end up being gooped! LOL! 😂

Educator Skona Brittain withholds the rules for even longer than me. She lets the students discover them by repeated explosions. I must try this!

Issai Schur was a Russian mathematician who studied at the University of Berlin under Ferdinand Frobenius (whose coin problem could easily have been added to this collection). As a secular Jew in Nazi Germany, Issai was stripped of all academic titles; he escaped to Palestine in 1939, only to die in poverty.

Bubbling Cauldrons is inspired by his work.

I love doing this activity around Halloween — in fact, I bought small plastic cauldrons just to use for it. I've done it successfully with a wide range of kids—from 1st grade to 8th grade. The basic puzzle involves arithmetic practice as well as logical thinking. And the extensions to other sequences reveal interesting number-theoretic properties. As with virtually all of Gord's puzzles, this one allows for kids to create their own versions by modifying the sequences.

Skona Brittain became an entrepreneur with her whimsical mathematical clocks. She jumped into home schooling with her children Shelly and Sandy, and now runs an eccentric math circle, the Santa Barbara Math Ellipse.

This Infinite Puzzle that keeps on giving was first explored in 1916 by mathematician Issai Schur. When a pair of students have found a way to get eight frogs in the two cauldrons (that's as good as they can get; nine frogs is impossible), just give them one of the following:

- Starting with 1, get as many ODD frogs as possible into two cauldrons.
- Get as many EVEN frogs as possible into two cauldrons. As always, you must start with the smallest even number and work your way up... 2, 4, 6, 8, 10, 12, 14, 16, etc. The best you can do is 16. (Is it coincidence that 16 is double the 8 frogs found in the original puzzle?)
- Get as many frogs as possible into three cauldrons. The best possible answer is below. I have not yet had a grade-two student find this during class.
- A magical golden cauldron exists. Throw as few frogs as possible into it. Numbers in the gold cauldron do not explode. Start with two iron cauldrons and a gold cauldron. Numbers with the same remainder when divided by five must go into the same cauldron. This extension and the next one are for older students. They are from the Julia Robinson Mathematics Festival.
- Using all iron cauldrons, choose squares, cubes or numbers from the Fibonacci sequence. Again the objective is to try to get as many of these into a certain number of cauldrons.
- Using just two iron cauldrons, how high can you get if a cauldron explodes only when three frogs in the cauldron sum to the frog being dropped in?
- What if a cauldron explodes only if exactly one pair of frogs sums to the frog being dropped in? If two pairs of frogs both sum to the frog being dropped in, nothing happens.

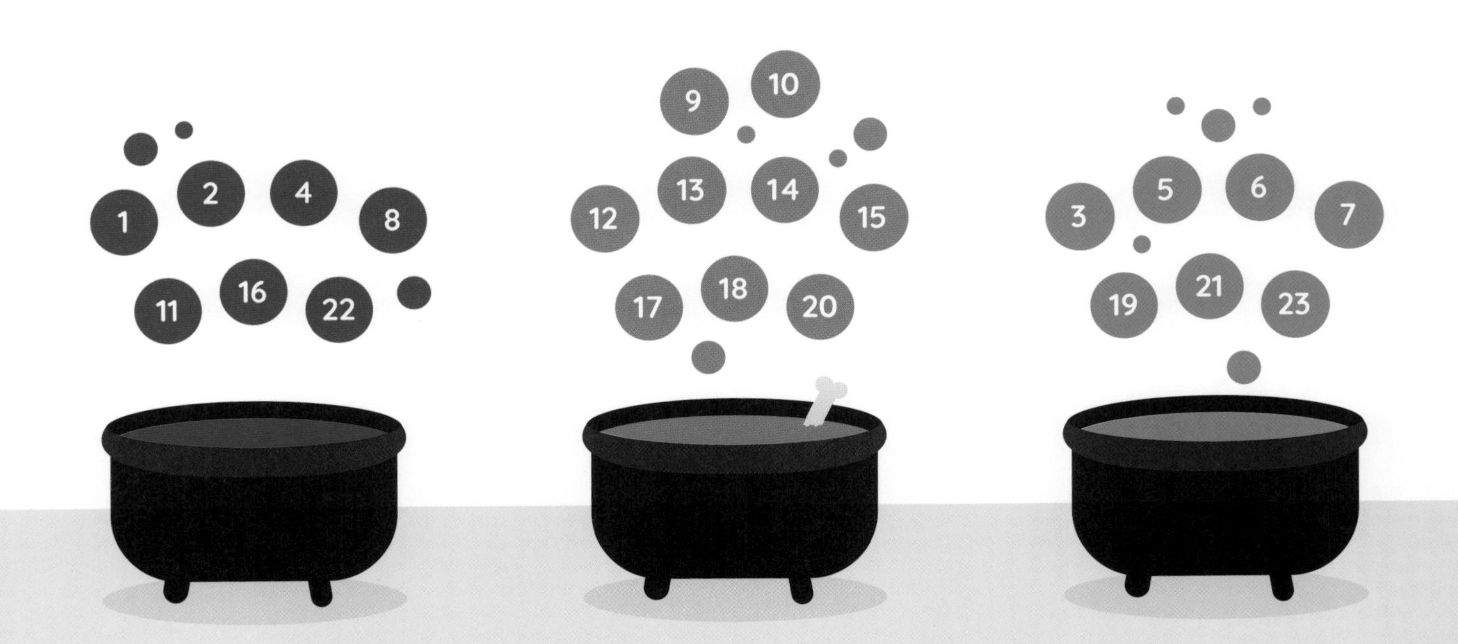

BANYAN TREE

A lonely banyan tree pushes its roots into the ground year after year. Here are snap-shots after ten, twenty and thirty years.

How do the roots grow? There are simple rules to discover. What is the next number to grow under the root 30-20-14-10-6-4-3-2-1?

SPOILER ALERT!

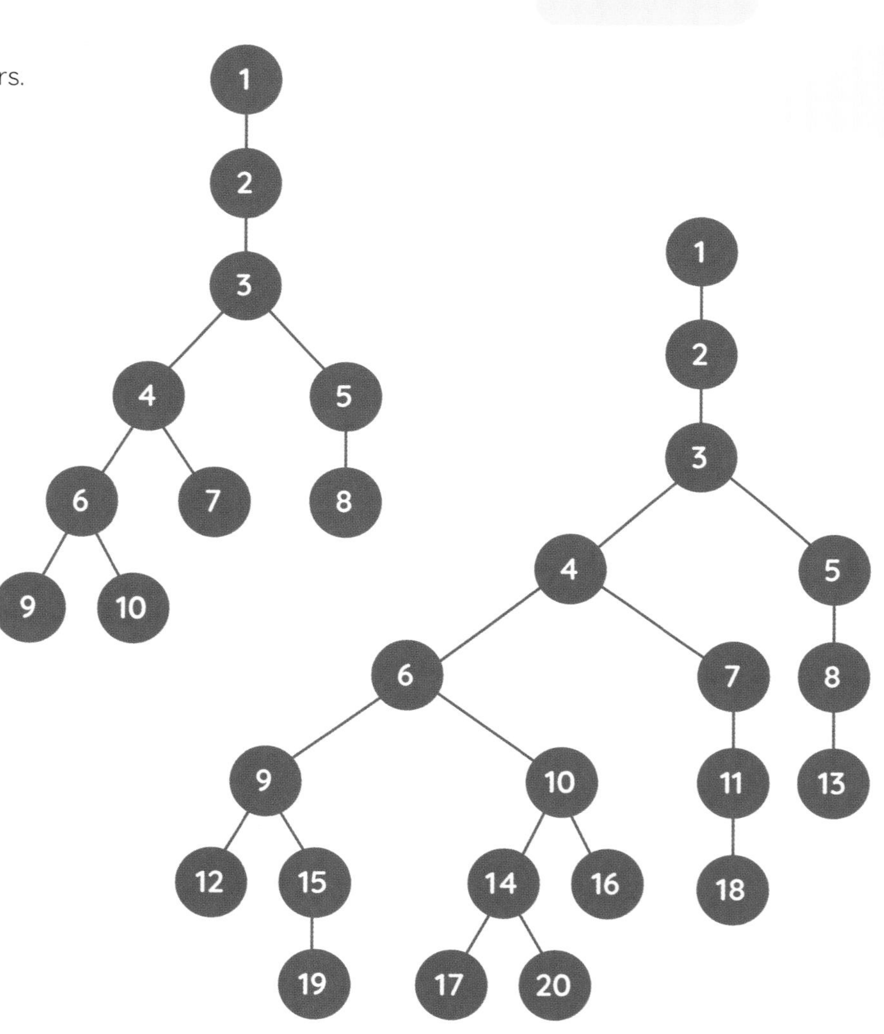

720

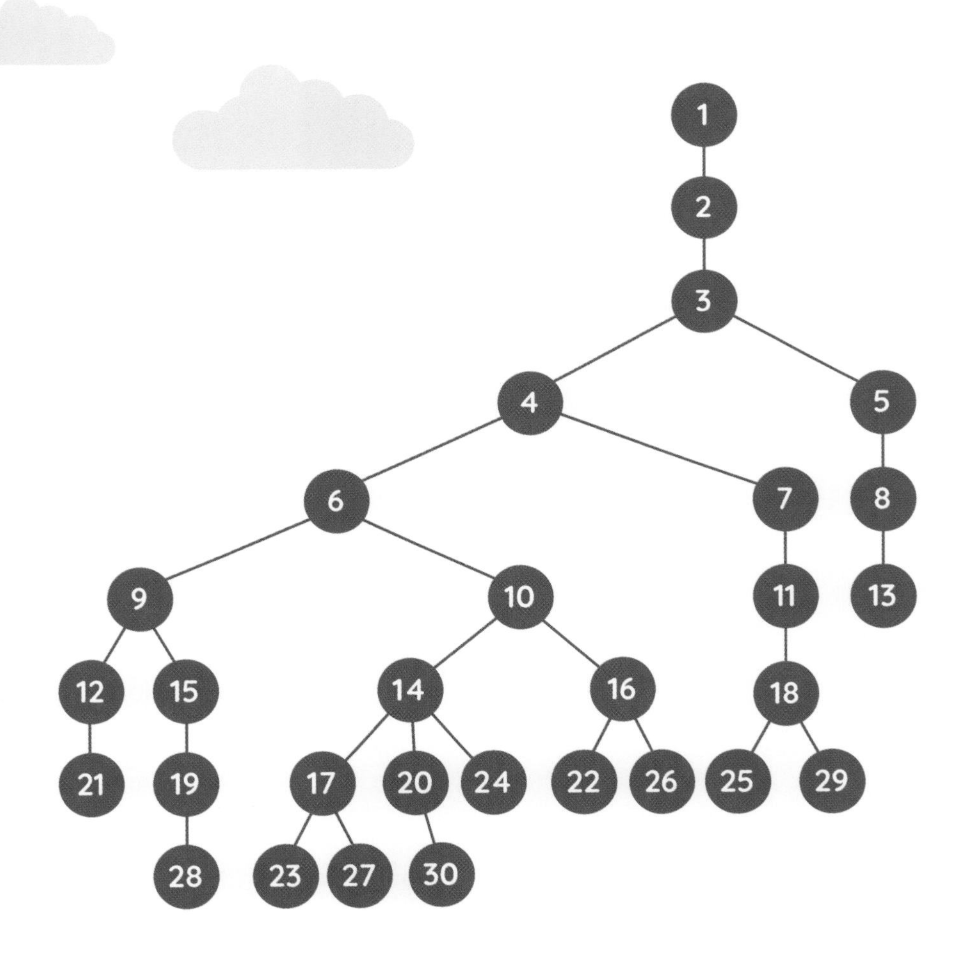

In the classroom these roots should be revealed one integer at a time, asking kids to:

- predict where the next integer might appear.
- make observations.
- hypothesize about the rules.
- kill hypotheses that don't fit the observations.

This is the scientific method. Ironically, the scientific method is best first taught not in science class by sending kids out into nature to get their hands dirty, but in math class with mini-mathematical universes like these banyan trees. Why? Because the rules of our real universe are complex and difficult to discover, whereas the rules of mini-mathematical universes are designed to be discovered.

In class there should be as many wrong hypotheses as right hypotheses. That is the best indicator that children are feeling open enough to talk about their ideas. If you are not getting enough wrong hypotheses, specifically ask for some. "What's a hypothesis that we know is wrong?"

Here are the same roots at age 60. Some observations:

- Going down each root, the integers get bigger.
- Each integer 1-60 appears exactly once.
- Some roots look like they are only going to have even numbers.
- The left bit (6-9-12-21-33-39) is forgetfully skip counting by threes.
- The numbers spread out horizontally as the roots age. The 9 and 10 used to be close together.
- Some roots like 13 and 24 look like they have stoped growing.

Here are some rules that describe the lonely banyan tree's growth after 2:

- An integer grows on a root where it is the sum of two different numbers. Example: The 14 grows on the root **10**-6-**4**-3-2-1 because 14=10+4.
- If there is a choice, it grows on the smallest possible number. Example: The 14 does not grow on the root **11**-7-4-**3**-2-1 because it prefers to grow under the smaller number 10.
- Growth under a number increases left to right. Example: under 31 there is 35-45-48.
- Numbers are midway between numbers underneath them. Example: 14 is neither far left over the 17 nor is it far right over 24. It is central.
- The lowest numbers on the roots are equally spaced left to right: 39-47-49-53-51-60-56-57-55-58-44-52-59-48-50-24-54-38-42-36-43-29-13.
- Roots with the same number of numbers go to the same depth.

The lonely banyan tree has chaotic roots. Other banyan trees have rules that lead to repetitive patterns.

Can you figure out the rules for this close cousin of the lonely banyan tree?

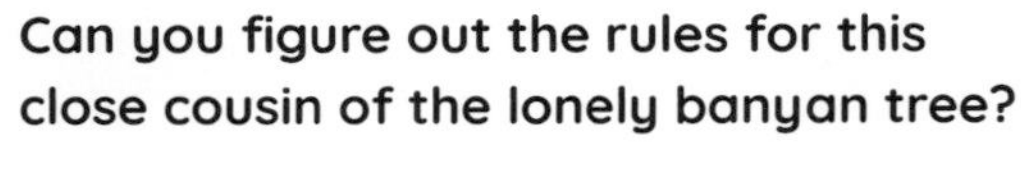

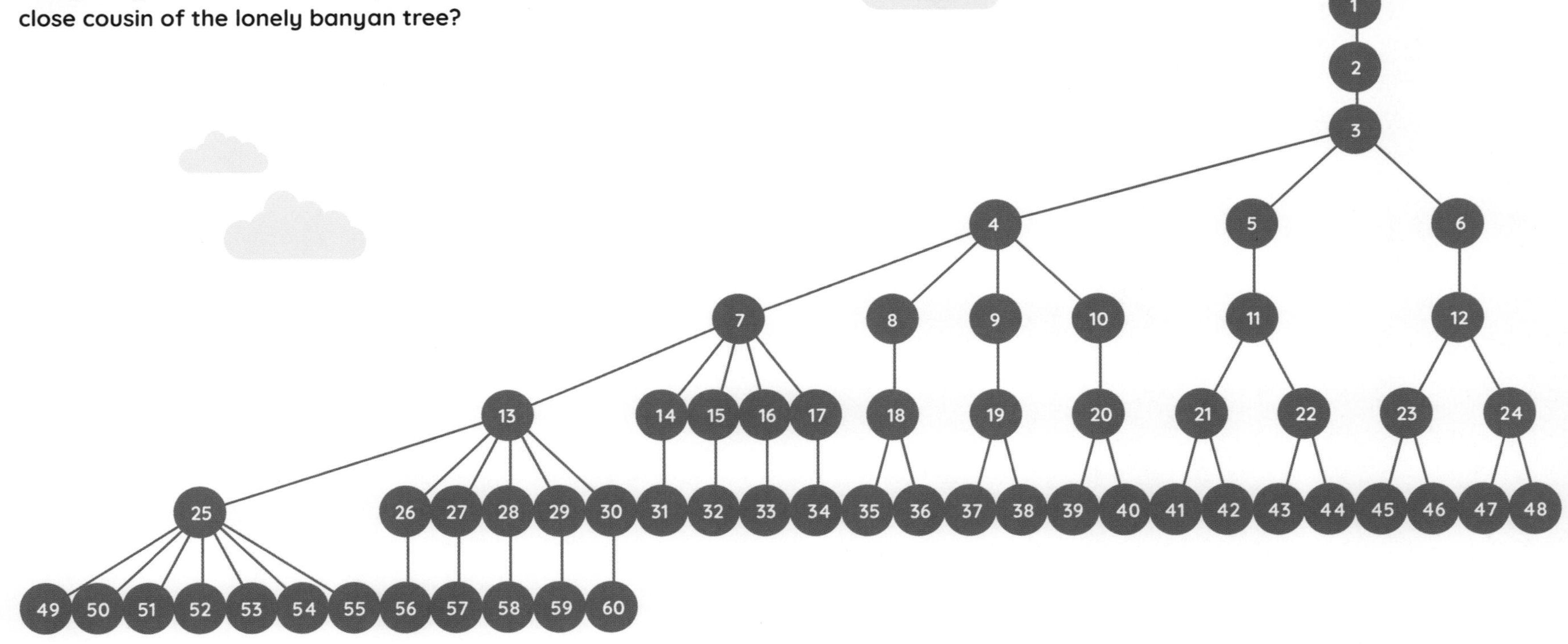

The rules are identical to the lonely banyan tree except for the first rule. Instead of summing exactly two different integers, sum two or more integers. Example: The 15 grows on the root 7-4-3-2-1 because 15=7+4+3+1.

Enough of these solitary trees standing alone! It's now time for us to explore the Infinite Forest!

Here are the roots of two sixty-year-old trees in the Infinite Forest. All roots in the Infinite Forest grow according to the same rules. Can you figure out what they are?

SPOILER ALERT!

Here are the most critical rules in the Infinite Forest:

- All roots start with 1.
- Each year, the next integer grows on a root where it is the sum of two numbers, but unlike our previous trees, these integers can be the same. Example: 32 grows on the root 16-8-4-2-1 because 32=16+16.
- Each growth must be on a root that is as short as possible. Example: In the lower tree, 17 grows on a length-5 root: 16-8-4-2-1 (17=16+1) rather than a length-6 root: 15-10-5-3-2-1 (17=15+2). Often there is a choice of many equally short roots. It is because of this choice that there are an infinite number of trees in the Infinite Forest.

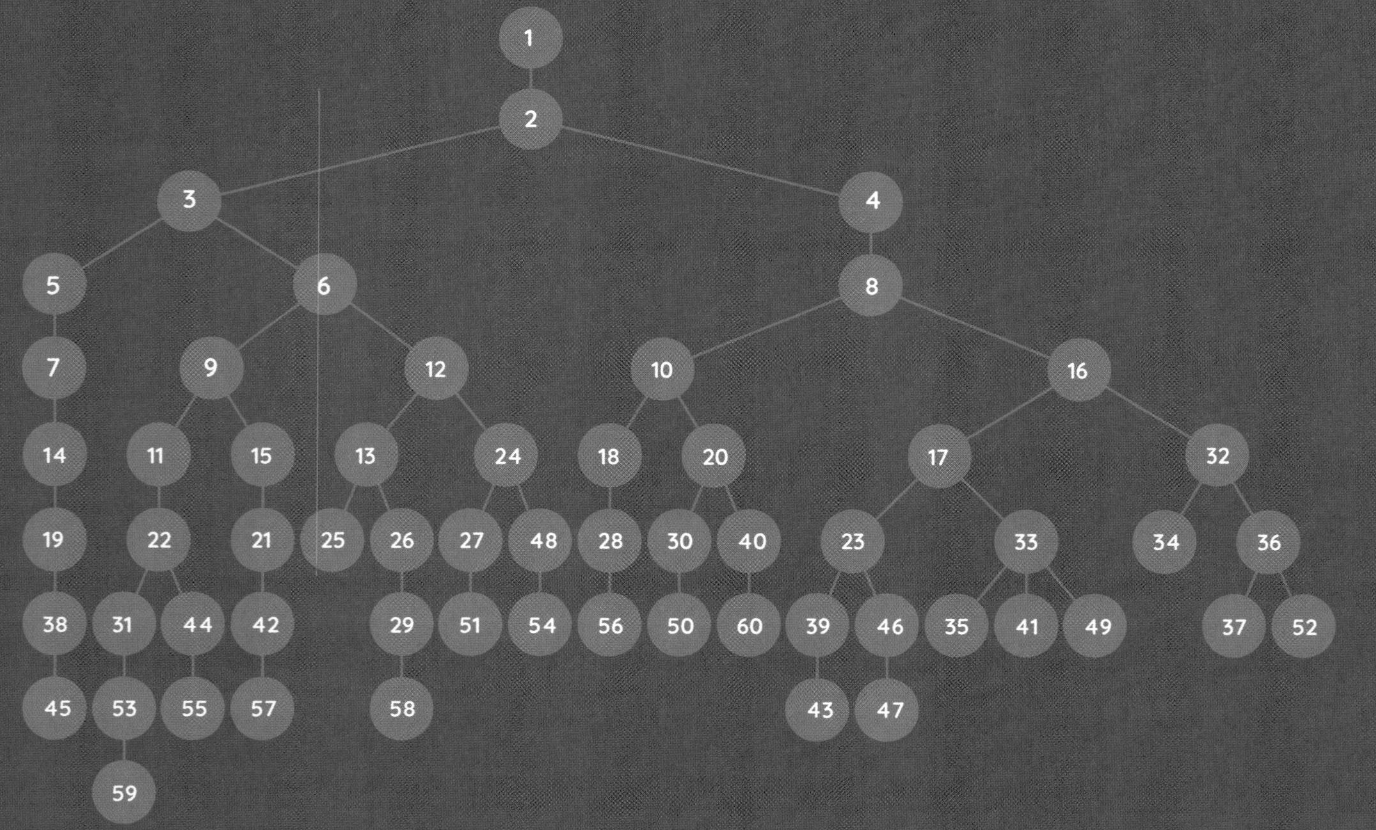

Trees in the Infinite Forest fear their roots going too deep. For some numbers there is no choice. The number 6 is always at the same root-depth. It is either on the 4-depth root 6-3-2-1 (6=3+3) as above or the 4-depth root 6-4-2-1 (6=4+2) as on the left.

What is the smallest number that can be at two different root depths?

SPOILER ALERT!

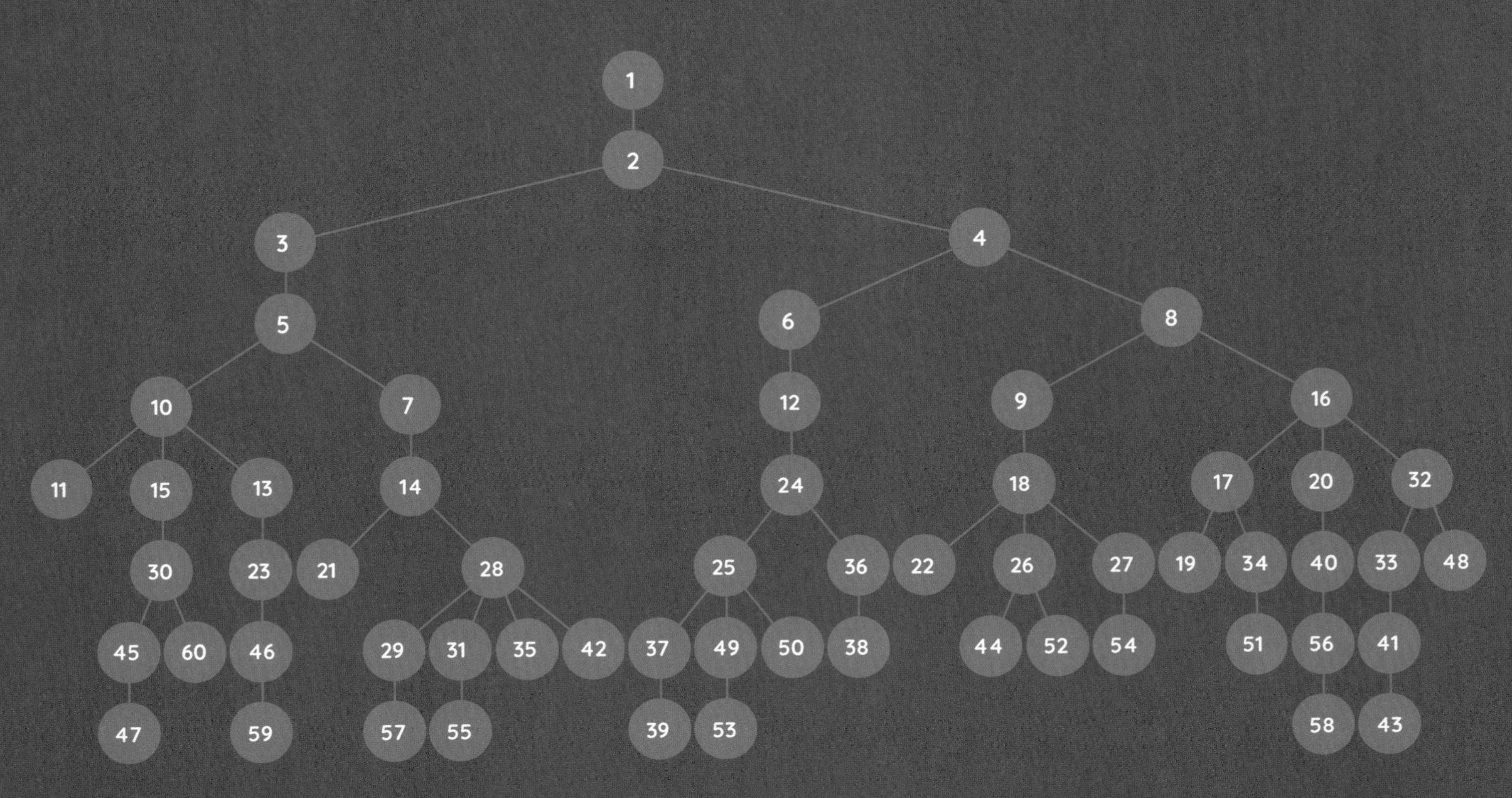

The smallest number that can be at two different root depths is 15. It can be part of a 6-depth root like 15-9-6-3-2-1 or part of a 7-depth root like 15-11-10-6-4-2-1.

I don't know the lowest number that can be at three or four different depths.

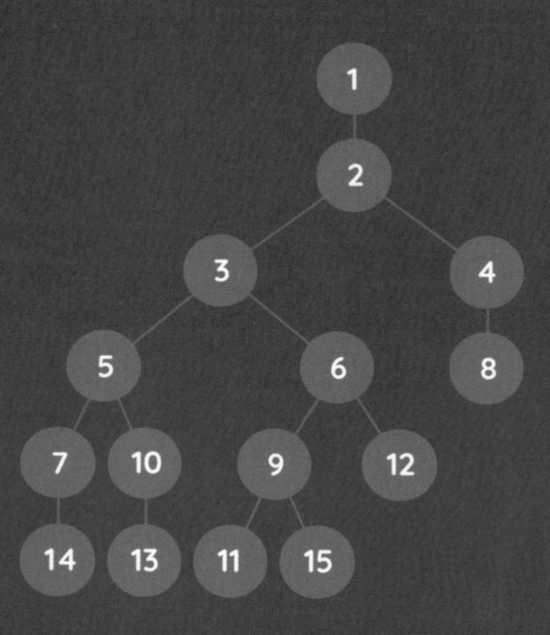

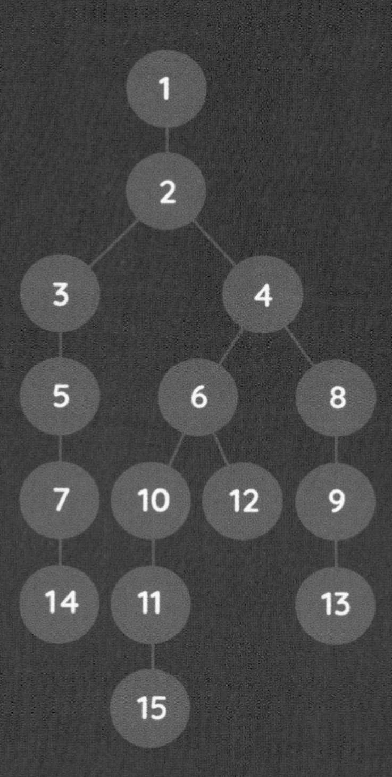

A competition between two infinitely old trees is decided by looking for the smallest positive integer that is deeper on one tree than the other. The loser is the tree that has the deeper one. **There is one infinitely old tree that never loses a competition. It exists, but its branching roots remain a mystery.**

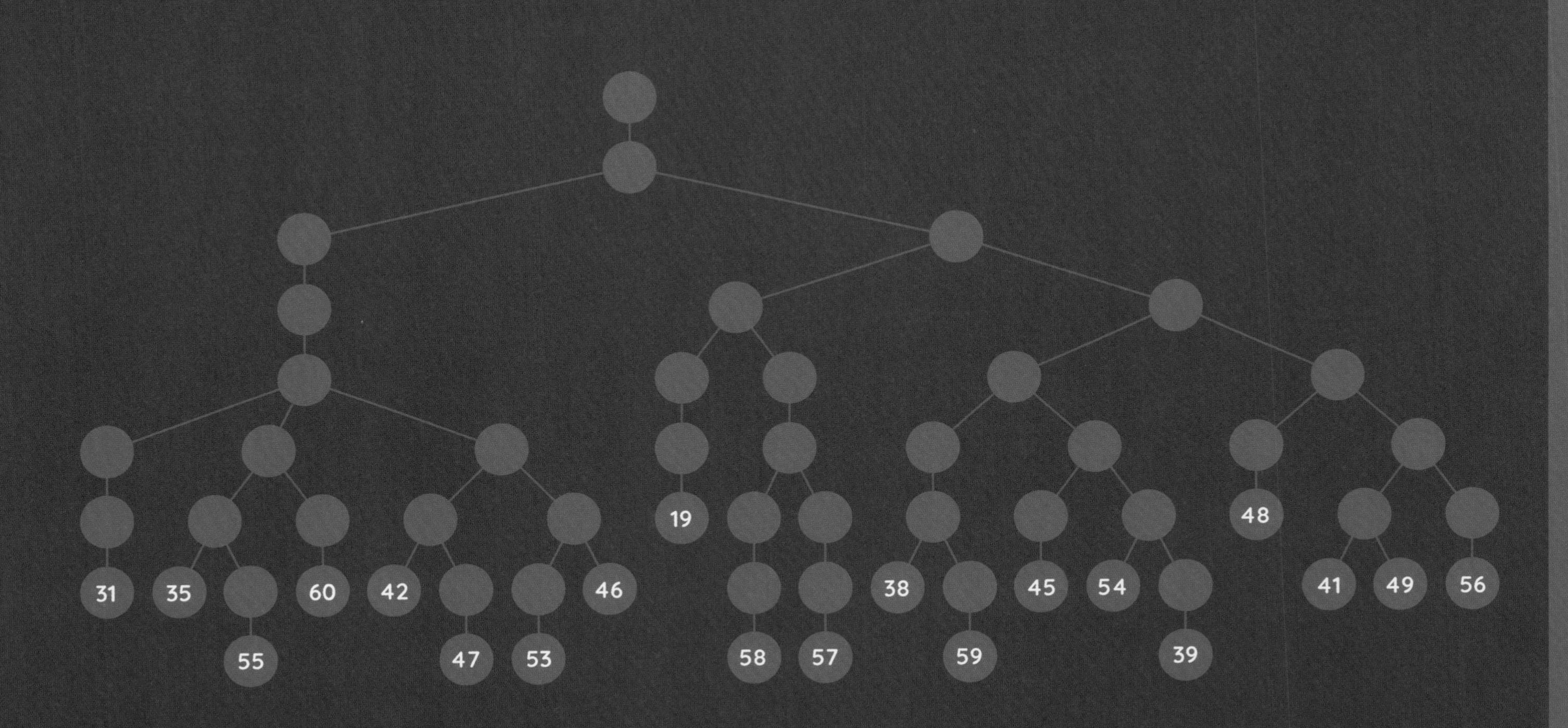

This puzzle has the bottom number of each root given as a hint. Do all such puzzles in the infinite forest have a unique tree as the solution? Are trees of infinite age possible to solve? I don't know the answer to either question.

The solution to this puzzle is on the next page.
SPOILER ALERT!

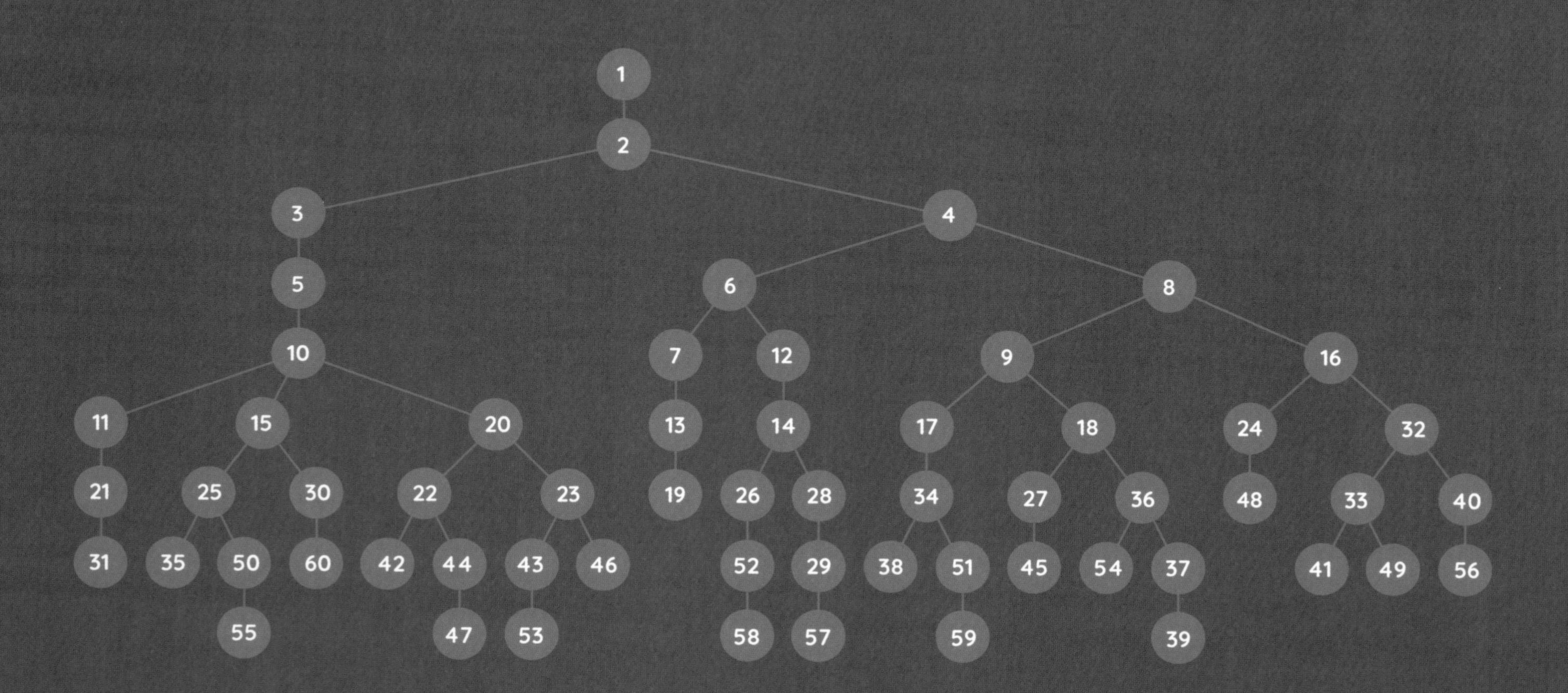
1
2
3
4
5
6
8
10
7
12
9
16
11
15
20
13
14
17
18
24
32
21
25
30
22
23
19
26
28
34
27
36
48
33
40
31
35
50
60
42
44
43
46
52
29
38
51
45
54
37
41
49
56
55
47
53
58
57
59
39

Hi!

I'm Gordon Hamilton. I'm the father of two teenagers, but professionally best known as the inventor of board games like Santorini and as the director of MathPickle.com

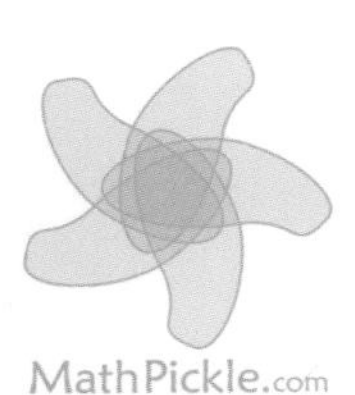

Eighteen years ago my son was just about to be born and I thought I would see what education had in store for him. I volunteered as the token mathematician at a math fair in a local school in Calgary, Canada.

I wandered around the exhibits solving puzzles and enjoying the students. Then it happened. A grade-five girl explained her puzzle. It was well known, but it was new to me. I tried to solve it this way. I tried to solve it that way. She knew I was the mathematician. The more I struggled, the broader her evil little grin became.

I didn't solve the puzzle till I got home that evening, but on leaving the school I realized that my failing was the most wonderful gift I could have given that girl.

MathPickle and this book are a direct result of that encounter.

The 14 Infinite Pickles in this book are designed to engage a wide spectrum of students. They are all original. I'm really proud of them. 🙂

- Teaching is an experimental science.
- Arithmetic is overemphasized. Speed is over-emphasized. Slow, tough problem solving is where we need to spend the bulk of classroom time.
- **Mathematics must be hard!** The primary purpose of mathematics education is to teach students to think. If someone says they'll make math "easy", run away—they're missing the whole point.
- **Why we teach mathematics?** #1 reason: We teach mathematics because of what remains when the mathematics is forgotten.
- Try teaching rigorous thinking with a backdrop of religion, politics, literature or even science. It doesn't work. **Choose math.**
- If I have to choose between a grade 6 child who can recite their timetables quickly and a child who can think sharply enough to beat me in a board game, I'll choose the child who can think. Hopefully I don't have to choose.
- **Give the gift of failure.**
- It is in the repeated daily exposure to failure that students lose the stigma of failure and are able to fully engage.
- Parents often request that gifted students be accelerated through the curriculum. These students hop, skip and jump from success to boring success until they hit university—then splat! Failure needs to be built into the classroom experiences of all children—especially gifted children—starting in kindergarten.
- Physical Education and Math are the only two subjects that give a crisp feedback loop for the educator. Students either dunk the basketball or miss. The answer is either right or wrong. This crisp feedback makes a perfect medium to teach students how to move and think.
- Some educators emphasize "real-world" problems too much. "Real-world" is irrelevant. The only relevant question when determining if a curricular problem is good: **Does it engage the full spectrum of student ability?**
- **Don't Interrupt!** Math class is ending in 5 minutes. 85% of students are engaged. Is it time for reflection? Probably not.
- In the same way as we don't have a class called "vocabulary," we should not have a class called "mathematics."
- Rename elementary mathematics classes "problem solving" classes. These classes would use mathematics to solve interesting puzzles.
- If a student doesn't show their work, stop complaining and give them a tougher problem.
- Ask an impossible problem every day. If your students can't trust you to give them nice, respectable problems, math class becomes unpredictably delicious.
- Don't wrap-up math in nice, neat packages. **End with a question.** "I wonder if we can make a Venn diagram with four circles?"
- Your number one objective in creating a lesson is not to make it **SIMPLE** but to make it **ENGAGING.**
- The scientific method should first be taught in the mathematics classroom. Students should guess the rules of a Mini Mathematical Universe that was created to reveal its secrets. The real world's laws are too complex.

- Rules are boring. Do not start an elementary school puzzle by explaining the rules. That will bore 20% of the students. Instead, **start with an emotional experience...** Get the class to try to solve the puzzle without knowing the rules. Of course they fail. Laugh together! How tongue-in-cheek "nasty" of you! After they fail, tell them one rule—or let them guess at the rule that made them fail.
- Pairing off students to solve problems together is worth experimenting with. Many children find social math engaging.
- **Beauty over truth.** There is no excuse for ugly mathematics in elementary school. It should look beautiful. Too often we celebrate worksheets that are all true and boring and ugly.
- The puzzles in this book are not intended for the elite few. Problem solving is at the heart of a quality mathematics education for everyone.
- **#1 job of parents:** Beat their child in a board game every week. That's how parents best complement the problem solving classroom.

ULAM numbers

The page numbers in this book are Ulam numbers. Read about them on page 6.

1	53	145	243	382	502	673	847
2	57	148	253	390	522	685	849
3	62	155	258	400	524	688	861
4	69	175	260	402	544	690	864
6	72	177	273	409	546	695	866
8	77	180	282	412	566	720	891
11	82	182	309	414	568	722	893
13	87	189	316	429	585	732	905
16	97	197	319	431	602	734	927
18	99	206	324	434	605	739	949
26	102	209	339	441	607	751	983
28	106	219	341	451	612	781	986
36	114	221	356	456	624	783	991
38	126	236	358	483	627	798	~~999~~
47	131	238	363	485	646	800	...
48	138	241	370	497	668	820	

A puzzle collection of infinite delight!

David Martin President of the Math Council
Alberta Teachers Association

The Infinite Pickle **does what Martin Gardner did, but widening the door to include much younger learners.**

Scott Kim Puzzle Master

The Infinite Pickle **should be in every math teacher's back pocket.**

Stephanie Englehaupt Elementary School Teacher

This book was written and illustrated by humans.